Excursions
in and about
Newfoundland

During the Years 1839 and 1840

VOLUME 2

JOSEPH BEETE JUKES

CAMBRIDGE
UNIVERSITY PRESS

CAMBRIDGE UNIVERSITY PRESS

Cambridge, New York, Melbourne, Madrid, Cape Town,
Singapore, São Paolo, Delhi, Tokyo, Mexico City

Published in the United States of America by Cambridge University Press, New York

www.cambridge.org
Information on this title: www.cambridge.org/9781108030908

© in this compilation Cambridge University Press 2011

This edition first published 1842
This digitally printed version 2011

ISBN 978-1-108-03090-8 Paperback

CAMBRIDGE LIBRARY COLLECTION

Books of enduring scholarly value

Travel and Exploration

The history of travel writing dates back to the Bible, Caesar, the Vikings and the Crusaders, and its many themes include war, trade, science and recreation. Explorers from Columbus to Cook charted lands not previously visited by Western travellers, and were followed by merchants, missionaries, and colonists, who wrote accounts of their experiences. The development of steam power in the nineteenth century provided opportunities for increasing numbers of 'ordinary' people to travel further, more economically, and more safely, and resulted in great enthusiasm for travel writing among the reading public. Works included in this series range from first-hand descriptions of previously unrecorded places, to literary accounts of the strange habits of foreigners, to examples of the burgeoning numbers of guidebooks produced to satisfy the needs of a new kind of traveller - the tourist.

Excursions in and about Newfoundland

Joseph Beete Jukes (1811–1869) was a geologist who studied at Cambridge under the famous Adam Sedgwick (1785–1873) and eventually became a prominent member of the Geological Survey of Great Britain. In 1839, after many field expeditions in England, he was appointed to a survey of Newfoundland, a place about which he had until then been in 'utter ignorance'. The explorers failed to find the hoped-for mineral wealth they had been sent to prospect for, and returned to Britain. In 1841 Jukes joined the H.M.S. *Fly* as a naturalist for an upcoming expedition to chart the coasts of Australia and New Guinea. The *Fly* set sail for the Pacific in 1842, the year in which this two-volume account of Jukes' Newfoundland experiences was published. Volume 2 focuses mainly on Jukes' scientific observations, and includes descriptions of the island's natural history, geography and geology.

Cambridge University Press has long been a pioneer in the reissuing of out-of-print titles from its own backlist, producing digital reprints of books that are still sought after by scholars and students but could not be reprinted economically using traditional technology. The Cambridge Library Collection extends this activity to a wider range of books which are still of importance to researchers and professionals, either for the source material they contain, or as landmarks in the history of their academic discipline.

Drawing from the world-renowned collections in the Cambridge University Library, and guided by the advice of experts in each subject area, Cambridge University Press is using state-of-the-art scanning machines in its own Printing House to capture the content of each book selected for inclusion. The files are processed to give a consistently clear, crisp image, and the books finished to the high quality standard for which the Press is recognised around the world. The latest print-on-demand technology ensures that the books will remain available indefinitely, and that orders for single or multiple copies can quickly be supplied.

The Cambridge Library Collection will bring back to life books of enduring scholarly value (including out-of-copyright works originally issued by other publishers) across a wide range of disciplines in the humanities and social sciences and in science and technology.

EXCURSIONS

IN AND ABOUT

NEWFOUNDLAND,

DURING THE YEARS 1839 AND 1840.

By J. B. JUKES, M.A., F.G.S., F.C.P.S.;

OF ST. JOHN'S COLLEGE, CAMBRIDGE;

LATE GEOLOGICAL SURVEYOR OF NEWFOUNDLAND.

IN TWO VOLUMES.

VOLUME II.

LONDON:

JOHN MURRAY, ALBEMARLE STREET.

1842.

CONTENTS OF VOLUME II.

CHAPTER IX.

CHAPTER X.

CHAPTER XI.

EXCURSIONS

IN AND ABOUT

NEWFOUNDLAND.

CHAPTER IX.

Difficulties with the House of Assembly—Excursion to
Topsail and Conception Bay—Start for St. Mary's Bay—
The Butterpots Hill—Holyrood Pond—St. Mary's—
Father D—— and his Farm—A Wedding—Proceed to
Colinet—Visit the Rocky River and return Overland to
Placentia—Excursions to the North-east Mountain, to
Meraskeen, and the Ragged Islands—Cross from Long
Harbour to Chapel Arm, and return to St. John's.

On returning from the ice, I found that in
consequence of some squabbles between the
House of Assembly and the Council, the
Money Bill as first voted by the House had
been thrown out by the Council, and that consequently the grant of 600*l.* for the geological
survey was become a nonentity. In the next
Money Bill the House inserted no grant for
the survey, but voted me 100*l.* to pay my

passage home.   I was considerably amused at
this, but determined on taking the 100*l.* and
pleasing myself about going home.   I was
resolved, at all events, to see the rest of
Avalon, and complete as far as I could the
survey of that part of the country.   However,
after some time they determined to vote 450*l.*,
350*l.* as salary and 100*l.* for travelling expenses, and a member called on me privately,
recommending me not to take a vessel, since
in that case I could make as much money
out of this grant as out of the former one!
I need hardly say that my object was not to
make money, but a geological survey; however, as they had shown so strong a disposition
to throw me overboard, without notice given
or reason assigned, I determined to take them
at their word.   My travelling expenses the
previous year, including the cost of the vessel,
had amounted to upwards of 400*l.*   It was
now near the middle of May, much valuable
time was lost, and if I stayed to equip a small
craft, I should not be off before June: so I
determined to trust to chance for my passage
alongshore, and to set off with only a servant,
a knapsack, and a shooting-jacket.

Meantime, on April 25th, Professor Stuwitz
and I set out on an excursion of a couple of
days to Topsail and the shore of Conception
Bay. For the first five or six miles the road
was finished, and was in a condition good
enough for a horse to trot along it. We then
deviated to the top of Branscombe Hill, and
I had the pleasure of introducing Stuwitz to a
Newfoundland wood, which considerably astonished him, though it was not so bad as many
I had seen. Beyond this the road crossed a
marsh for about three miles, and was only
marked out, without being gravelled. There
was a pretty pond or two, with a brook running,
out to St. John's Harbour. A wooden house,
or tilt, was occasionally to be seen on the side
of the road, and the first rude approach to
agriculture was observable in small gardens or
rough little fields fenced with logs. We then
entered a wood through which the road had
been cut, although the stumps and roots of the
trees had been left. In this wood was a large
pond on our left, closely girt with the dense
forest, from which a brook sprang out, leaping
down a wild and rocky channel into the little
narrow valley that leads to Topsail. We

crossed this brook by a wooden bridge, and presently came to a place where three men were employed in blasting and clearing away a mass of rock that obstructed the road. The rock was an intensely hard, close, tough, siliceous stone with few joints, and was about as troublesome a piece of stuff to cut through as I had ever seen. Although it was so early in the season, we were attacked by mosquitoes whilst talking to these men, and received one or two bites. These mosquitoes were of the kind called in Newfoundland *gallinippers :* they are a species of gnat, with long thin legs and a slender body, having a long trunk or proboscis, the end of which they insert into the skin, and suck the blood till their bodies increase to three or four times their original size, becoming quite red and bloated. A swelling forms round the bite, which continues painful and irritable at times for a fortnight or three weeks. Another kind, which in Newfoundland is always distinguished as the mosquito, is a little black fly with white thighs, and is more like a small house-fly than a gnat : and this insect seems to bite off a piece of the skin, as the wound bleeds copiously, and a small cicatrix forms, which is very teasing and irritable. A third kind is the sand-fly, very minute and troublesome, and found principally on the sea-beach near wood and running water. The mosquitoes of the second kind dance around you in thousands in the woods and marshes during the day, but are quiet at night; while the gallinipper with his sonorous hum hovers over you both by day and night, sailing about, banishing all sleep from your eyes, and alighting now and then on some exposed part, when you suddenly feel the sharp tingling of his bite.

The descent into Topsail is very pretty, the last two miles of road being finished and gravelled over. It winds along one side of a ravine, with a brook below and rocks and woods about it, until Conception Bay appears in sight, with some flat land and a great pebble beach stretching alongshore to the south. We arrived at four o'clock in the afternoon, and put up at a house which, although not a regular inn, was considered the travellers' house, and was almost the only one in which a bed was to be had. The people were very neat and clean, and gave us a dinner and supper of salt fish, eggs, and tea. The woman of the house had a bad foot, and she told us that the surgeons of St. John's charged 5*l.* for a visit to this place, distant about twelve miles, while it was impossible for her to get to St. John's without sailing down to Portugal Cove or round Cape St. Francis, either way making a two days' journey of it at least. I mention these facts as showing the difficulties of overland travelling in Newfoundland, even where roads are partially complete.

April 26th.—In setting out this morning for Broad Cove, our first obstruction was a brook cutting through the beach, and crossed by a bridge consisting of one thin slippery pole: as I had a barometer on my back, I preferred wading through, to the chance, or, I may say in my case, the certainty of falling off the pole into the brook. We then clambered on along a rough beach, and climbed up a cliff into the wood. We here found a tolerable footpath leading through the wood along the edge of the cliff on our left, 80 or 100 feet above the sea, and having a bold precipitous hill called Topsail Head, rising immediately on our right to the height of 400 or 500 feet. From this path we had many highly picturesque and beautiful views of the bay with its three islands

and its bold and varied headlands and rocky
shores.   Beyond Topsail Head were several
small clearings, and some comfortable cottages, with little rocky coves and rough
wooden platforms and stages along the face of
the cliff, made accessible by rude stairs   At
Broad Cove we had again to cross a brook on
a single pole, which we did with the assistance
of another pole held across by a man as a kind
of rail. Here we got some dinner, consisting of
fresh herrings, tea, and bread and butter. We
then climbed the cliffs at the back of the cove,
following a track which had been cleared of
bushes, and was intended to be a road, though
now consisting of a mere succession of crags and
great boulders.   At the top of this rocky bank,
which was about 200 feet high, was a cleared
space, from which the view of Conception Bay
was the finest and most beautiful we had yet
seen.   A tolerable track was then found along
the new line of road, and the brooks were
crossed by bridges till we came to the borders
of Twenty-mile Pond : after this we found the
road gravelled, and shortly reached the Portugal Cove road, and so returned to St. John's.

Having at length come to a right under-
standing as to the continuance of the survey

for the present year, Professor Stuwitz, who
had taken my old vessel, the Beaufort, into his
service, offered me a passage in her as far as
St. Mary's Bay or Trepassée.

May 19th.—We set sail with a north-north-
west wind, and a jumping sea, which before
we got fairly round Cape Spear sent both of
us into our berths. By the time we reached
Ferryland Head, however, I was better: here
the wind shifted into the west, and as this
would be dead against us when we got to Cape
Race, we beat up into the harbour of Aquafort,
and anchored near its head. This is a long inlet, with pleasant shores, and cliffs rising to a
height of about 200 feet. It takes its name of
Aquafort from a pretty cascade on the northern
side, where a brook shoots its waters over a
cliff into the sea.

May 20th.—Fine morning, with light, variable winds. I determined here, if I could get a
guide, to walk across the country to St. Mary's
Bay. I consulted accordingly with a gentleman,
the only resident merchant in Aquafort, who
recommended me to go to a man in Ferryland,
as the person best acquainted with the interior
of the country. Stuwitz and I accordingly
walked across to Ferryland by a very fair road

between three and four miles long. We there
found the man who professed to be a " pilot "
for the country, but on consulting him he said
he would not undertake it without 5l. for himself, and 3l. for another man to carry provisions,
&c., and that it would take us three days of
hard work to get to St. Mary's. I was very
angry at first at what I considered a rascally
attempt at extortion ; but on relating it to a
gentleman of Ferryland, he did not seem to
think the demand an extravagant one, more
especially at this season of the year, when the
people were busy preparing for and commencing the fishery. The man, moreover, by no
means seemed willing to undertake the job,
and absolutely refused to abate a jot, so I gave
it up. I was still, however, determined to go
to the Butterpots and the range of hills running a few miles from the coast, and made inquiries for a man to take me that far, but could
not find one willing to undertake it. One or
two said it was very difficult to go in and come
out on the Aquafort side, and recommended
me to go to Renews and try it from there.
There was a road made as far as Renews, and
I therefore determined to do so.

May 21st. — After some delay in getting breakfast, I set off accompanied by Simon, at 7 A.M. for Renews. The road was pretty tolerable, passing over the high grounds between Aquafort and Fermouse, sweeping round the head of that long and picturesque inlet, and then over high ground again to Renews. The distance from Aquafort to Renews was about seven miles, and we caught occasional glimpses of the Butterpots Hill over the woods on our right, and were evidently nearest to it when passing round Fermouse harbour. After some little trouble, we found a house at Renews, where they undertook to give me a bed for the night, and the next thing was to procure a "pilot" or guide. Luckily we met with some of our ice comrades, and the brother of one of them, named Tom Coady, offered to go with us. At 10 A.M. we set out, walking through a brook at the head of the harbour, the water of which was sufficiently cold to the feet. The day was fine, and the sun shone on the side of the hill we were going to visit, which, indeed, seemed so near us, that the fuss made about going to it would have appeared to a stranger perfectly ridicu-

lous. It was certainly not more than six or
seven miles in a straight line. We struck along
a small path into a wood, but in about a mile
and a half this ended, and we came out on a
marsh. Here we sat down to rest, when Bell
pointed at a little distance, and before I could
get up, a brace of ptarmigan rose, of which I
bagged one; we then toiled across the marsh
till twelve o'clock, when I began to feel almost
knocked up. We accordingly halted, lighted
a fire and made some tea, which, together with
some beef we had brought with us, gave us
strength for a fresh start. At one, after traversing some more marsh, we came on some small
rocky " barrens," where I killed another brace
of ptarmigan. We were then obliged to strike
into the woods, and soon afterwards came down
to the main brook, the same we had crossed near
the harbour; wading through this, we had then
some more very thick scrubby wood to scramble
through for some distance, when we came out
on some marshes and clear spaces near the foot
of the first ascent of the hill. An amphitheatre
of dark craggy precipices stretched away to the
west at the distance of about a mile; but before
us was a gentle ascent, which appeared to lead

by an easy route and a succession of open slopes
and short cliffs to the summit of the hill. We
pushed on accordingly to the first exposed sheet
of rock, which I found to consist of beds of
common slate, dipping from the hill at a considerable angle. Climbing up this, we found
ourselves on a bare table of rock tolerably smooth
and easy to traverse, but had hardly proceeded
300 yards when to our great disgust we found
ourselves on an insulated eminence, and that a
valley 200 feet deep and a quarter of a mile
wide, filled with the densest possible mass of
wood, still intervened between us and the principal body of the hill. I descended the side of
the hill with all care on account of the barometer
slung across my shoulder, but notwithstanding
my precaution, a thick sheet of moss gave
way, and sliding with me along the smooth
face of the slate-rock, down I went into some
stunted fir-trees below, being partly caught
and brought up by the barometer-case. Pushing our way through the woods, we arrived
at the next slope, which was so rocky and
precipitous, that we were obliged to pass my
dog Bell from one to another as we climbed
up. Most provokingly, after surmounting this

eminence and walking across its tabular summit, we again found our route interrupted
by a valley full of wood precisely similar to
the last. There was now, however, no time
to be lost; so we clambered down, scrambled through the wood, and again climbed
up the opposing precipice by the help of the
trees that grew in the clefts, to a still loftier
eminence than before. Even this we found
cut off by a third valley, from the summit of
the hill, but it was a much smaller and shallower one than before, containing merely a
small pond and a bed of " tucking bushes."
These consist of a kind of dwarf juniper or other
fir-tree, with very thick short stumps and strong
flat interlacing branches. They grow breast-
high, and are so close, firm, and level at top,
that in some places a man can stand on them.
As, however, it is not quite possible to walk
everywhere along their tops, it is necessary
sometimes to wade through them, and where it
is not possible to avoid them by going round,
I think 200 yards of " tucking bushes " in an
hour would be quick work, and certainly much
severer toil than six miles of plain ground.
By walking in the water of the shallow edge of

the little pond, we evaded these bushes, and after another short stoppage for a slight refreshment, were soon afterwards on the top of the hill, which was one rounded mass of bare rock, something like a rude dome. All the principal mass of the hill, after the first eminence, consisted of porphyry, passing here and there into sienite. On opening my barometer, I found the tube broken and the mercury gone, so I was left to guess at the height of the hill, which I should give at about 1100 feet. The view from the summit was very commanding, extending from the ridge at the Bay of Bulls to the country west of Trepassée Bay. The whole range of the eastern coast was visible, but in the west and north-west the view was shut in by rising ground, over which, however, I thought I could just discern the top of the north-east mountain of Placentia. In the direction of Conception Bay was a line of inconceivably broken and rugged country, hardly constituting a distinct ridge, but covered with knobs and hummocks. Some more decided hills with a steep face to the east were called " Bread and Cheese," " Bold Face," &c., and I thought I could make out the Cats Cove hills,

bearing exactly magnetic north by my prismatic compass, but could not distinguish,
among the many rugged eminences, which belonged to the Butterpots of Holyrood.   It
seemed to me inconceivable how the common
notion could have originated that the hill seen
from Conception Bay, and called the Butterpots there, was the same hill as this Butterpots near Renews; and yet such was the
universal belief among the inhabitants. We
counted from the summit of this hill eighty
ponds or lakes of water, many of which were
two or three miles across, and none less than
one hundred yards. There was a great assemblage of large ponds in the direction of Trepassée, from which flow two streams, namely,
the main brook into the harbour of Renews
and the Black River into Biscay Bay. The
wind was blowing, and it was piercingly cold
on the top of the hill, but we stayed there an
hour while I took a round of angles and
bearings with the prismatic compass and box
sextant, and left, completely chilled, at about
five o'clock. On coming down we selected
an easier route than the one by which we
went up, keeping more to the eastward,

along the top of a regular ridge, and having only one precipitous face to descend. On reaching the bottom of the hill, we found some dry barrens with open ground, and a grove of birch-trees close by; and while stopping to eat some biscuit and drink some sugar and water, we consulted as to whether we should sleep where we were, or try to get to Renews before dark. Tom Coady, however, said he knew an easier route, avoiding the brook altogether, and coming into the road between Renews and Fermouse; and, as I had no blanket, I determined to try. The way was almost entirely through marshes, and very wet; we were frequently nearly up to our knees in the moss, and the constant wet and cold chilled my legs and feet so much that I could hardly walk. At one place I stumbled over a great boulder and strained my left knee. By perseverance, however, and dogged labour, we succeeded in reaching the road about eight o'clock, just as it was getting dark. Even Bell was dreadfully tired; and as we sat down to get breath on the road side, she began scratching up the moss, and making a nest or form among the brush-wood, evidently with the hope of passing the night there. I think it is in 'Woodstock' that Scott notices the habit dogs have of turning round before they lie down; and he makes old Sir Henry Lee say he cannot account for it. Once before, after a hard day's shooting in England, I had observed a spaniel make a nest for himself in the thick grass and brambles on the side of a wood, which he did by scratching away the sticks from a particular spot, and turning round several times with his body bent, till he formed a snug circular form, in which he lay down with a defence from the wind all round him. It is interesting thus to see on the rug before our parlour fires the habits and instincts displayed that were intended for a wild and savage life. Rousing up Bell from her bivouac, we again addressed ourselves to the road; and as we really had one now to avail ourselves of, we reached our quarters about nine o'clock. I found a neat little room upstairs, and tea, eggs, fish, and fresh-baked cakes, the master of the house coming to sit with me as a point of courtesy, and apologising for his absence if he left. There was also a clean bed, which, soon after a hearty meal, I laid myself down in.

May 22nd.—A sharp frost this morning, and I congratulated myself on our having pushed on last night. Stuwitz had agreed, if a favourable wind sprang up, to set sail and call for me at Renews; but as there was now a breeze blowing from the south, I set off back again for Aquafort. The air was quite clear, and it was very hot along the road in the country, but we could see a fog-bank out at sea; and on getting down to Aquafort we found a bitterly cold wind and a thick fog driving up the harbour. The coldness of the wind, which was now east, gave evidence of there being ice on the banks in considerable quantity drifting down from the north, a circumstance that frequently occurs at this season of the year. After sitting writing and reading for four or five hours, I found on getting up my left knee intolerably stiff and painful, but hoped it would recover again with exercise. The chief resident here, Mr. W———, kindly invited us ashore in the evening, and procured some scallops and other sea animals for us from the sand of the harbour.

May 23rd.—The wind came up from the north-east; so we beat out of the harbour and ran round to Trepassée, anchoring at that place at night.

May 24th.—My knee pained me during the night, however I determined to walk across to St. Mary's from this place; so after taking leaving of Stuwitz, I went in search of a man to act as pilot, and help Simon to carry my knapsack, and our provisions and implements. Simon had provided himself in St. John's with a tin kettle divided into two compartments, which could on occasion be separated, and in one of which we could boil our tea, and in the other a ptarmigan, or any other eatable. This sufficed for our cooking apparatus; a couple of hatchets provided us with beds, and sometimes with shelter, but we were obliged to carry a change of clothes and shoes, and my note-books, ammunition, hammers and hammer-bag, blanket, mackintosh, and various other matters. I soon found a strapping young fellow called Con Kennedy. His hand had been blown off by the bursting of a gun, but his shoulders and legs were as broad and as strong as ever, and he could still shoot, resting his gun across the middle of his left arm. He immediately agreed to go with us to St. Mary's for a pound and a pair of shoes, as his own would be worn out by the journey. As I had no shoes to spare, however, I offered to give him thirty shillings, and let him find his own shoes, which he accepted. We were put across the harbour in a punt, and followed a little path through a thin skirt of wood to the barrens. My leg pained me at first, and made me quite lame, bringing me to a slow march the whole of the way. The country was chiefly barren, generally clear of woods, and having only a few marshes and ponds interspersed about. It was by far the easiest country to traverse I had yet seen; in proof of which we met, to my utter asto-nishment, a man on horseback. It was the first time I had seen such a phenomenon, except on a regularly made road. It was an old man on a strong pony : he could never go beyond a walk, and the pony was plastered with mud up to the flaps of the saddle, in passing through some of the soft places, but still he was going down to Trepassée, having come from Peter's River, in St. Mary's Bay,

and this was the only space of ground of that
extent that I saw anywhere in Newfoundland
where such a thing was possible without a
regularly constructed road.   About half way
to Peter's River we crossed a brook which runs
into a pond about a mile long, and afterwards
through another pond lower down, emptying
itself into the sea near St. Shots. In the afternoon we came to the very pretty valley called
Peter's River, with flat meadow-land on each
side of a considerable brook, bounded by
abrupt ascents of rock and scrubby wood. We
got some refreshment at the first house we
came to, and then walked to Holyrood Pond,
the path conducting us sometimes along a
sand-beach, and sometimes along the edge of
the cliffs.   This pond was once an inlet of the
sea, communicating with a little bay, the
western point of which is called Cape English.
The tide and current which sweep up the bay
along this shore, aided by the prevalence of
south-westerly gales, have drifted a beach of
large pebbles and boulders to the shores of the
little bay, and have completely blocked up the
mouth of the inlet, separating it from the sea
by a bank of pebbles two miles in length, one

hundred yards across, and from ten to twenty feet in height. This beach, however, was now intersected at the end next us by a passage one hundred yards wide, through which a rapid current was pouring; so we hailed and fired a shot, and presently a four-oared punt came across the pond and took us up to the house. The people here treated us very kindly and hospitably, giving us the usual fare of the country. At night they showed me into a small bed-room, and the good woman hoped I should have no objection to let the young man who came with me (my servant Simon) have part of my bed. As he was a very clean decent lad, I assented to this, but his modesty preferred the kitchen floor, so he did not make his appearance.

May 25th.—Sailed in the punt about thirteen miles along Holyrood pond to the end of a road that goes down to St. Mary's harbour. The pond is twenty-five miles long, and about two miles wide. It is a fine sheet of water, but the ground about it is rather tame and uninteresting. In autumn all communication between the pond and the sea is prevented by the pebbles swept in by the

south-west gales and heavy swell.   During
winter the water accumulates owing to the
brooks which run into it, until on the melting
of the snows in spring it stands almost as
high as the top of the pebble-beach, and oozes
through the upper pebbles.   The inhabitants
assemble at the beginning of June with their
shovels, and cut a small trench through the
beach ; the water then rushes out, widening
the channel, and the pond falls to the level of
the sea, and afterwards rises and falls with the
tide, which rushes in and out with great rapidity.   Herrings, cod-fish, salmon, capelin, and
seals then all enter, the water becoming quite
salt, and sea-weed growing on the banks below
high-water mark.   The banks are enclosed
when the winter gales repair the breach, and
herrings and cod-fish are taken all the year
round in the pond.   In winter the pond becomes merely brackish, and trout from the
neighbouring brooks are found in it.   Before
the country was inhabited, the beaver and the
otter lived on its banks : deer are plentiful still,
and from various causes the remains of these
animals, and those of the wolf, the fox, and the
bear, must frequently be deposited at the bot-

tom of this piece of water. In the mud, therefore, at the bottom, there is probably a curious
admixture of land and fresh-water as well as
marine plants, and of marine and fresh-water
shells, while the bones of the trout are mingled
with those of the cod, the herring, and the
shark, and the remains of seals, beavers, and
land animals are confusedly jumbled together.
Were such an accumulation of fossils found in
the mud or silt in the valley of a river, how
puzzling would it appear, and yet there is no
improbability in the idea of this pond being
some day elevated into a dry valley, with a
brook running along the bottom, winding
through beds of mud, sand, or clay, containing all the things mentioned above, and possibly others. It is, moreover, curious that the
pond, a little way inside the pebble-beach, is
much deeper than the sea outside for some
miles. The average depth of the pond is
from thirty to fifty fathoms, while the people
assured me there was scarce half that depth for
five or six miles round Cape English. Were
the country elevated, therefore, till the bottom
of the sea became dry land, Holyrood pond
would be a lake of entirely fresh water, and

to this alternation and combination of fresh water with marine beds would succeed others purely fresh water. The probability of such things having frequently happened is, undoubtedly, very great: but it is worth while to notice the circumstances under which they really do take place at present; and we may see, from this example, how very complicated may be the results of one simple action—as for instance, of elevation.

From the middle of the western bank of the pond, a road about three miles in length, and nearly finished, leads to St. Mary's — a scattered place, with few good houses, and a harbour, neither very safe nor commodious. We put up at the house of Mr. Ferrars, which, though not an inn, was the place where strangers usually resorted to; and he sold provisions and rum. We found everything in confusion, the only room he had besides the kitchen being filled with bags of bread and flour. At night, I learnt that they had but one bed to spare, and it was fully expected that I should admit Simon to a share of it. As I could not condemn the poor fellow to the kitchen-floor for an unknown number of nights, I even conformed to the custom of the country at once, and insisted on his taking half the bed, though with much reluctance and many apologies on his part.

May 26th, 27th, 28th.—Detained in St. Mary's. My knee was too stiff and painful both day and night to enable me to move about much ; and the only boat I could hire was one going to the head of the bay, which did not get off till the 29th. The people were all very busy in getting ready for the fishery : the boats had returned from St. John's and other places, bringing the summer supplies of bread, butter, pork, nets, lines, hooks, and clothes, and also rum. Some one had, moreover, been to St. Peter's and brought some brandy : the consequence was, that there was a good deal of drunkenness and some fighting ; and St. Mary's struck me as being the most rough and disorderly place I had yet seen. The stipendiary magistrate usually residing here was now absent: this and the abundance of spirits were probably the causes why things presented so unfavourable an aspect. About Cape English they fish chiefly with nets, the water there being shoal enough to use them : but in other places they use the hook and line, running

out to Cape St. Mary or Cape Pine in
their half-decked boats. All the while I was
in St. Mary's, there was a thin blue haze
overhead, partially obscuring the sun: there
was no smell of smoke, but it was said to
proceed from a great fire in the woods on the
other side of the bay. Father D. invited me
to tea with him one evening, and took me over
the farm he was trying to establish round his
house. The country around slopes gently
from the interior to the water, the land rising
only to a moderate height of about 200 feet:
the subsoil generally consists of gravel, and it
is altogether better adapted for agriculture than
the neighbourhood of St. John's. In two years
Father D. has cleared and reduced to a tolerable state of cultivation a considerable space
of ground, probably twenty acres, and he in-
tends to grow oats, barley, and turnips. He
has several cows, horses, pigs, and sheep, all
very fine of their kind. His sheep were certainly the fattest and best-looking, and had the
most wool of any I had seen in the country.
It was now shearing time, and the wool appeared very fine, and was used by my enter-
tainer for his own dress, such as stockings, &c.

At present all his cattle ramble about loose in
the woods; and in the winter are sometimes ha-
rassed and destroyed by the wolves. He hopes
to have a complete farm cleared and fenced in
a few years; but told me he thought it would
require ten years before he could realize the
money expended, and begin to gain a profit. I
accompanied him in the evening to a wedding
at a fisherman's house close by. The company, consisting of young men and women
with a few of the seniors, assembled about nine
o'clock, and sat round the room drinking grog
for some time. Presently a fifer struck up a
tune, and reels and jigs began. There was much
bustling of women in and out of a back room;
and about half past ten the bride was said to
be ready, the music ceased, a table was brought
forward, the priest put on his scapular, and the
bride and bridegroom came forward and knelt
before the table. The priest opened his book,
asked them the usual questions in English, and
rapidly read a number of Latin prayers, and
the ceremony was completed. Instantly there
was a struggle among the young men around
for the first kiss; but as the bride was not yet
off her knees, and the bridegroom was kneeling

beside her, he had the best chance, and won accordingly. A plate of cake was next produced and put upon the table : the bride and bridegroom came forward, took a piece of cake, and deposited several dollars on the table before the priest. Each one in the room then came forward and took a piece of cake, leaving a dollar before the priest, who stood behind the table thanking them as they came up. I had nothing but two-dollar notes in my pocket, of which I laid down one, when Father D. took it up, looked at it, shook his head, and said, " Ah, sir, upon my word, that's too much !   A five pound note !" I was immediately held in great honour accordingly ; and as I could not contradict the priest before his flock, I was obliged to receive the acknowledgments of the people with the best grace I could. He told me afterwards this was his object, and that the people would now do anything for me. When all had made their contributions, they stood round while he counted the money and declared its amount, stating it, if I recollect rightly, at something under 10*l.*, which with my 5*l.*, as he was pleased to call it, made nearly 15*l.* " Wish it was more for your sake, sir," was the universal response round the room. Supper was now brought in, consisting of tea and hot cakes; after which, there was more dancing and grog-drinking, nearly to the break of day.

May 29th.—The two men, John Davis and old Joe, to whom the boat belonged in which I intended to proceed, reported her ready this morning, but the wind was against us. At two o'clock, however, it seemed inclined to favour us, and we set off. The boat was a small unpainted skiff, with two old sails and a couple of clumsy oars; and besides the two men, Simon, and myself, there was a Mrs. Quigley, of Harrycove, near Colinet, as a passenger. A light wind carried us out of the harbour, but it soon fell calm; and a light breeze from the north-west sprang up dead against us. We then toiled slowly on with the oars till dusk, when, just as we entered the passage between Colinet Island and the main, the wind shifted into the south-east, and we sailed down to Admiral's Beach, where we anchored, and went ashore at a lonely house. The people here were very civil, and received us most hospitably, giving us an excellent supper of fresh cod-fish, which, as I had lived lately on herrings, was quite a luxury. They gave up their bed to me, likewise, which had not only blankets, but sheets, another luxury; and I was heartily glad I had left St. Mary's. There was a light flame to-night on the other side of the bay, distinctly visible, and evidently proceeding from a great conflagration, which they said had been raging in the woods for three weeks. Bright patches of glancing aurora were likewise flickering all over the northern hemisphere, but without any distinct form.

May 30th.—Sailed early in the morning with a fine south-west breeze to the mouth of Salmonier, where I had intended to land; but finding nothing but the common clay-slate, I determined to push on for Colinet. We entered a narrow inlet between an island and the main, where we landed our female passenger, and then pushed on in very shoal water into the Tickle. What the origin of this word Tickle may be I am at a loss to conjecture; but it is applied all over Newfoundland to a narrow passage or strait between two islands, or other points of land. This Tickle

is so shoal in some places that it is almost dry at
low water ; and once or twice we grounded, and
Davis was obliged to get out and shove off. We
soon, however, got round the island into the
main arm again, and then sailed up to the
head of the inlet, and the mouth of the Colinet
River.   Here we grounded, and flung out our
anchor, and on hailing stoutly, Davis's wife
came off in a small leaky punt, and took us
ashore.   His house was clean and comfortable,
and situated in a small valley, down which a
little brook ran into the river.   He had several
acres of meadow-land, only part of which had
required clearing, and two or three potato-gar-
dens, and also two milch cows and a bullock ;
but a winter or two ago had lost a flock
of nine sheep and three cows by the wolves,
who came two nights successively, and tore
them in pieces, although they were in the immediate neighbourhood of the house.   He has
a salmon-weir close by in the river, which is
about sixty yards wide, but shallow and rapid :
the weir he leaves to the care of his wife, a
servant, and old Joe, (who lives in the same
house,) while he himself fishes for cod in the bay

during the season. Old Joe had been a hunter
or trapper, till the beaver and other fur-bear-
ing animals were nearly all destroyed. He
then carried letters and messages across the
interior of the country during the winter, and
acted as "pilot" to any one wishing to cross
with him. He was now, however, getting old
and stiff, and lived generally with Davis. The
latter had no grant of his land, or other title
than that of occupancy ; but as it was a fertile
and pretty little spot, I recommended him
by all means to obtain a title. The wolves
were still troublesome, three having been shot
during the last winter at Harrycove between
Colinet and Salmonier.

May 31st.—After breakfast this morning,
there being no meat or fish in the house, we
determined to separate into two parties, and
try the two rivers which empty themselves
here, in search of some food. Simon, and
Davis with his gun, accordingly went up the
Colinet River, while I accompanied old Joe to
the banks of the Rocky River. The entrance to
the Rocky River is broad and deep, with perpendicular cliffs of dark slate about twenty
feet high. It continued thus for half a mile,

obliging us to make a detour through the woods and across a small marsh, when we came out on what old Joe called "a very handsome fall." This fall consisted of two leaps, each about twenty feet high, with a foaming rapid between them, and a dark whirlpool below, the effect altogether being highly picturesque. Just above the fall, as we stepped out of the woods, old Joe drew me back and cried "Hist!" and immediately afterwards I heard the "couk" of a wild goose. Drawing back into the bush, I tried to steal through it and come down on them; the underwood, however, was so thick, and the fallen timber so plentiful, that they took the alarm. We then walked along the margin of the river, on a little rocky strand, where the footing was tolerably good. Half a mile beyond I spied two couple of geese, each with two young ones, little, callow, unfledged things. The old ones were very bold in the defence of their young, and allowed me to come within shot; but I found my light fowling-piece and small shot not heavy enough for the one I fired at. Joe, however, caught one of the goslings in the water; and we proceeded in pursuit of the

second pair, that had made all haste up the
river. Sending old Joe across and a little in
advance to drive them over to my side, I
again got within reach of an old one, and
firing both barrels together, I succeeded in
settling him, and Joe caught another young
one. Having got a dinner, we did not proceed further, as my knee became painful, but
came back with our spoil, Joe carrying the two
young ones in his cap. Simon and Davis came
back in the afternoon, having seen nothing
but one ptarmigan. The young geese immediately became tame and sociable, and a piece
of turf being cut they hopped upon it and
began pecking and eating, walking about the
kitchen as though they had been born in confinement. No bird seems more shy and wary
while wild, and none is more easily and quickly
familiarised than the Canada goose. They
will not breed, however, in confinement.

The view from the rising ground just north
of Davis's house was very pretty. Colinet
Arm was like a large lake, with low, woody
points projecting into it at intervals; and the
narrow entrance in the distance into St. Mary's
Bay presented the appearance of a river. The

distant ridges on each side, and the undulating grounds about, though not high or striking, were still very pretty, being richly covered with wood and spotted with green and yellow marshes.

June 1st.—We all made an expedition up the Rocky River, and went much farther than I had done yesterday. About three miles up, just at the place where we turned back yesterday, we found the fresh track of a wolf pointing down the river. We saw a good many geese, but they were now too wary: Davis got a shot with his long gun, but did no execution. The banks of the river are thickly wooded with spruce and birch, and the land generally seems of a much superior quality to that of the rest of Avalon. It is, moreover, low and sheltered; and if a road were formed along its banks to Trinity and Conception bays, it would offer by far the best locality for settling the interior of the country I had yet seen in Avalon. The river is shallow, but in wet seasons is navigable for little flat-bottomed boats, and when frozen in winter would make an excellent sleighing-road, being of an average width of sixty yards. In returning we left the river and walked through the country, where I shot a brace of ptarmigan.

June 2nd.—This morning I set out for Placentia overland, taking old Joe as a pilot. We walked down the landwash,* on the west side of Colinet Arm, for two or three miles, when we struck across the country over some rising ground to the head of North Harbour. On fording the river at the head of North Harbour we found the fresh track of a wolf. Just below we came to a house inhabited by two Irishmen, very civil and hospitable, but ragged and dirty. One of them, who went by the name of Big Tom, could hardly speak English ; and they both seemed regular wild Irishmen. They had abundance of good potato-ground cleared, and excellent potatoes, thirteen fine cows and six calves, besides six or eight very fine pigs, fed only on milk and potatoes. During the summer they generally caught 20*l.* or 30*l.* worth of salmon. With all this material for comfort, and with plenty of good wood which they might have for the

* This term is always used in Newfoundland for the margin of the sea, meaning that strip of land washed by the water.

trouble of cutting it, their house had no window in it, no floor but the ground, one stool and a log of wood; a board suspended from the wall and supported by one leg made the table; and two dark little holes partitioned off formed the bed-room and store-room. They seemed quite happy and contented notwithstanding, and apparently had no idea of comfort or cleanliness. The bed, which they gave up to Simon and me, consisted merely of a few blankets; and had not the night been so cold, I should have declined it. I took care not to undress, but notwithstanding this I sustained a sharp conflict during the night with prior occupants. The inhabitants of the harbour had suffered much the preceding winter from wolves, which had destroyed all the cattle belonging to some people a mile or two below; and one of the men said something had disturbed the cattle at daylight that morning, as they came running in towards the house bellowing and affrighted.

June 3rd.—Breakfasted at daybreak, and at five o'clock we set off, with a fine morning, and hoar-frost lying on the ground. For the first mile and a half after leaving the shore

we had to make our way up-hill through a
very thick wood, much tangled with old fallen
trees. There was a kind of path, however,
which enabled us to traverse it. On coming out
on the marshes at the top of the high ground,
we found them quite crisp with frost, and one
small pond with a coat of ice over it. At two
miles from the harbour we crossed a small
river flowing down to Cape Dog; and at three
miles or thereabouts we arrived at a considerable ridge, with a small peak, called North
Harbour Lookout. This ridge is a continuation
of the principal range of hills on this side of
Avalon, and we had Cape Dog and Mount
Sea-Pie on our left, and the South-east mountains, as they are called, on our right. Before
us stretched some level marshes, over which
rose a group of hills to the south-west called
Sawyer's Hills, and in the north-west a mass
of hilly and broken ground about Great and
Little Placentia. Behind us lay St. Mary's
Bay with its opposite shores. The ground of
this ridge had a singular appearance: it was
utterly bare, and the sharp edges of the thin
beds of slaty gritstone bristled up along it
almost like the edges of a set of knives. Pro-

ceeding on our route, we had then to cross, for
three or four miles, heavy marshes, and found
it very slow and toilsome work.   At one place
we came upon a wolf's track quite fresh, which
old Joe said had been made scarcely an hour
ago.   This was, probably, the fellow that had
disturbed the cows yesterday morning, and who
was still lurking about.   We looked about us
pretty sharply in going through the thickets,
hoping to get a shot, but saw no further
signs of him.   We crossed two brooks flowing
through the marshes down to Little Salmonier, and having at length got through the
marshes, we came to some low barrens and a
considerable pond with thin skirts of wood
about it.   Here, as we had now been walking
nearly six hours, I shot a ptarmigan, and
Simon plucked it as we went along.   When it
was ready we stopped under some bushes to
shelter us from the cold north-west wind,
made a fire, and cooked the bird and had
some tea.   We then proceeded through small
barrens interspersed with woods, little ponds,
and brooks or gullies.   I was still lame, but
my knee was much better, and I could keep
up with old Joe.   At length we got upon a

lofty barren, among some old burnt wood,
and came within sight of Placentia. On the
slope of the hill we found the road which is
intended to lead from Placentia to St. John's.
The line has been surveyed the whole way ;
and for the two or three miles nearest Placentia the woods have been cleared away, making
an opening twenty feet wide. Still the road
would have puzzled any one not accustomed to
the country, as in most places it was a mere
bed of boulders and crags of rock, with wet
boggy holes and soft places between them. To
traverse it required a succession of steps and
jumps from one slippery block to another, and
the wet places we had to wade through. This
delectable road conducted us to the head of
the south-east arm of Placentia, where we
found a house belonging to one Tom Kelly,
whose family immediately proceeded to supply
us with tea, and bread and butter. It was
now three o'clock in the afternoon, and we
had been ten hours walking about sixteen
miles, including two hours for stoppages : this
was at the rate of two miles an hour, which
for Newfoundland is pretty good work. Tom

Kelly now took us down the south-east arm to the harbour in his boat, where, to my great surprise, I found Stuwitz busily observing the amount of the dip or inclination of the needle. I got some very comfortable lodgings at the house of a Mrs. Morris, which, indeed, seemed to me replete with all possible luxuries and conveniences, after the rough work of the last fortnight.

June 4th to 8th.—I spent this time in examining the neighbourhood of Placentia, and in making preparations to proceed. I hired another man, a cousin of Simon's, named Tom Welsh, to assist in carrying our baggage,* but found no boats or vessels willing to give me a passage anywhere : they were all busy fishing, and nothing but the most extravagant prices would have induced them to disturb their arrangements. Mr. S——, the principal merchant, has a farm here, which is in better condition than any other in Newfoundland. The centre of the great beach of pebbles at the old mouth of the south-east arm is covered with

* His wages were to be 20l. for the summer, and he was to be found in shoes, besides board and lodging.

sand, and, at one place, is half a mile broad,
and sheltered by a bold hill immediately on the
east. This flat of sand has gradually become
covered with a rank marshy vegetation, which
has produced some vegetable mould. On this,
with great care and cultivation, are raised good
rye and barley, excellent potatoes, and a considerable quantity of grass: there are many
cows, and a regular dairy on a large scale; and
the dairyman assured me that during the summer months each cow yielded a firkin, or about
thirty-five pounds of butter a month. I was
unfortunate in not finding Mr. S——, but he
had not yet returned from Europe, where he
had spent the preceding winter. The weather
was now beginning to get very hot during the
day, and the capelin were coming in.

June 9th.—I had made arrangements once
or twice for a start into the North-east Mountain, where Tom Kelly had promised to pilot
me; and at last he came, about seven in the
morning of this day. We were, however, some
time before we could hire a punt to take us up
the north-east arm, but having succeeded in
that, we set off about nine o'clock. The north-
east arm is a fine inlet, about nine miles long,

and a mile wide, with bold cliffs and hills, rising
to the height of 400 or 500 feet. It expands
at the head into a shallow basin, containing
several islands, and receiving a considerable
brook. Making our punt secure, we struck
into the woods at the head of the arm, along
a path which Mr. S—— had had cut, and
which for the first mile was pretty good: it
was then very bad for half a mile, the woods
very thick, and clogged with many fallen trees,
which the people most inappropriately call
windfalls. The ascent was gradual, but the
woods were dreadfully close and hot. We then
came out upon some tolerable barrens, where
we had rather easy walking for a mile and a
half, gradually descending into a flat bottom,
and a marsh, which was intolerable: the mosquitoes also began to be very troublesome. At
the edge of this marsh we found the track
of a wolf, rather fresh, pointing towards Placentia; and we heard afterwards that during the
night he had killed five cows on the north side
of the north-east arm. Before we got across
this marsh we were obliged to stop and rest,
and drink some marsh water from one of the
numerous holes, while the mosquitoes slaked

their thirst a little on us. In walking they are
not quite so bad, but the moment one stops
they gather round in clouds, so that the hands
are incessantly employed in defending the face
from their attacks : in walking across the
marsh, however, the toil is too severe to enable
one to do more than occasionally raise the
hand to punish now and then one that is more
ferocious than the rest. Beyond the marsh
we came to a rocky ridge, which is considered
half way, and called, from a rude pile of stones,
" Naked-man Ridge."   From this point we
crossed from one rocky ridge to another, avoiding, as much as possible, the marshes and
the tucking bushes, till at half-past four o'clock
we found ourselves at the foot of a little hill,
covered with wood, in which was the tilt where
we were to pass the night. This tilt consisted
merely of a small, wooden, sloping roof, with
a gable at each end, having a hole at the
bottom of one end for a door, and at the top of
the other for a chimney. We were very glad,
however, to lay down our burdens and get
some tea. After resting an hour we set out
for the top of the hill dignified by the name of
the North-east Mountain : the ascent was gra-

dual, the hill being round-backed and flat-
topped, and in half an hour we reached the
top : the view was very beautiful. West of us
rose another similar hill, with a pond in the
hollow between us and it, a mile long, half a
mile broad, and surrounded with wood. Beyond the hill was a wide-spread heap of rugged irregular heights, forming the shores of
Placentia Bay, through the hollows of which
we could see the islands and the land on the
opposite side of the bay, and at one place the
water of the sea and a small vessel under sail.
North of us, at a distance of six or eight miles,
rose Spread Eagle Peak, beyond which lay
the water of Trinity Bay, the land about Dildo
Harbour, and the long headland of Tickle
Harbour Point: east of this, Spaniard's Bay,
Lookout, and Clarke's Hill, near Port de Grave,
closed in the near view; while still farther in
the same direction lay the Catscove Hills, the
Butterpots of Holyrood, the Flaky Downs, and
the long rugged ridge which stretches off thence
towards Renews. Between us and that ridge,
which was twenty or thirty miles distant, lay
a broad valley of comparatively low, but undulating ground, covered with a dense sea of

wood, through which the bright waters of a
pond peeped out at intervals. Numberless
other ponds, as well as many brooks, and the
waters of Rocky River, which was not more
than four miles distant, were concealed by the
woods. Towards the south this valley sloped
down to St. Mary's Bay, the whole expanse of
which lay stretched out before us with all its
headlands and islands, a most beautiful object.
The base of Cape English, down to the water-
line, was clearly visible, the horizon being some
miles beyond it.* To the west of the bay, and
due south of us, ran the continuation of the ridge
on which we stood, with Cap Hill and the south-
east mountains rising from it; and south-west
of us was a valley with a string of ponds, emptying themselves by a brook into the south-
east arm of Placentia. We counted in the whole
panorama sixty-seven ponds, none at a greater
distance than ten miles, and many more within
that space were hidden from us by the ine-
qualities of the ground. Although glowing hot
when we got to the top, we were cold enough
before I had finished taking angles and bear-

* This would make the height of the North-east Mountain
at least 1200 feet.

ings; and to keep themselves warm the men raised a pile of stones seven feet high, as a monument of our visit. On getting back to the tilt it was nearly dark, when we proceeded to cook a brace of ptarmigans I had shot as we came in, and cut boughs for our beds. By aid of some wet moss we made smoke enough to expel great part of the mosquitoes; and before we laid down I scrambled up the little hill at our backs, and smoked my pipe in silence on its summit. Before me, over the tops of the trees, rose the two hills we had just visited, bright in the clear moonshine, and every rock distinct, while in the valley between lay the little lake gleaming through the woods and winding round the flanks of the hills. About fifty feet below, the appearance of a few red sparks glancing occasionally among the dark woods from the chimney of our tilt was the only sign of the presence of man; and a leaf would have been heard to fall had it dropped within twenty yards of me. So pure and tranquil a scene of beauty, rugged as were its principal objects, it falls only to the lot of the traveller in lonely lands to behold.

July 10th.—I awoke once in the middle of

the night from the buzzing of a gallinipper, or
mosquito, and found that the fir-boughs on
which we lay had been placed too near the fire,
and that a flame was stealing quietly along
the leaves to where we slept, with our guns
and powder-horns beside us. The men jumped
up in a great hurry at my shout, and we soon
extinguished the cause of danger. At six
o'clock we had breakfasted, packed up our
traps, and were on our return. The morning
was overcast, and cooler and pleasanter than
yesterday, but my clothes were still saturated
with perspiration, making me feel very stiff
and uncomfortable, and my knee began to be
painful again: however, we kept on, only
stopping to shoot two or three ptarmigans,
and at eleven we reached our boat. The distance from the water to the North-east Mountain is about nine or ten miles, which again
gives two miles an hour as the average progress through this rough country. Those
who are used to it can do more, as I found
that men whom I could leave out of sight
when we were walking together on a made
road would have a corresponding advantage
over me in the wild country. On reaching

the water it began to rain, and before we got back to Placentia we had a regular soaking.

June 12th.—Stuwitz gave me a passage across to Merasheen, which we reached in a fog. After dinner, however, it cleared off, and we found ourselves in a snug harbour surrounded by rocky precipices and full of little rocky coves. There were several houses, but I went to Mr. K——'s, the principal merchant, to take up my own quarters, boarding my men with a planter, or fisherman.

June 13th to 19th.—Examined this island and its neighbourhood as far as weather and circumstances would permit. It is very long, narrow, and lofty, and about five miles from the harbour is a peak at least 600 feet high, where Captain Cook had a station when he surveyed this coast. The lofty parts are of course all barren, but there are several little spots alongshore at the foot of the hills where there are settlers cultivating gardens. Dr. M—— was living with Mr. K——, and was practising as a medical man in this thinly populated and scattered district. As most of his patients necessarily paid him in fish, he was

obliged to keep a man and a boat to go round
and collect his debts at the proper season.   He
kindly lent me his boat, and Mr. K—— accompanied me in it to Isle of Valen and the
Ragged Islands.   These islands form a singular and picturesque group of rocks along
the western shore of Merasheen, and we were
beginning a very pleasant cruise among them
when it began to rain; we then ran into a
small place called Merry Harbour (one of
the most dismal I ever saw), and went up to
the best of the two houses which composed
its habitations.   This, however, was but a
miserable hut, with half the roof burnt off,
and the people too careless to mend it.   As it
was occupied, moreover, by a man and his wife
and six children, we merely stopped to cook
some tea and fish, and pushed off for another
island, where was a winter-house at present
uninhabited.   This place was called " John the
Gong," which must be a corruption of some
French name, but what it was I did not make
out.   There was a snug little cove, in which
we moored the boat, and we found a tolerable house, consisting of one room in two compartments, in a little glen between two steep

hills covered with wood. The rain still continued: however, we made a blazing fire, and dried some fir-boughs before it for a bed. It was still rather damp than otherwise, and as we had brought no blanket we were awakened once or twice by the cold when the fire got low.

June 17th.—We set off at daylight and found the eggs of several sea-birds among the clefts of the rocks in these numerous islands, and then commenced our return, beating up home against a smart breeze. Although the sun was shining brightly we got very cold while sitting in the boat, and when about half way home we landed at Virgin's Cove, and sent the men on with the boat. Virgin's Cove is a small space of flat land beneath a bold cliff of 300 or 400 feet, over which, at one place, a large brook precipitated itself, and was lost in spray before it reached the bottom. The steep slope of the cliff was in one place rather more moderate and clothed with wood, through which a path led to the top. Following this, and crossing a mountain torrent on two poles laid over it side by side, we arrived at Captain Cook's station, at a spot called, from a tall pile of stones, "the Naked Man."

The view was bold and extensive, but the
land all along the west side of Placentia Bay
was barren and rugged in the extreme. We
walked down to Merasheen harbour by a very
rugged and precipitous track, and found the
sun, when we were sheltered from the wind,
intensely hot.

June 20th.—I had been detained by calms
and contrary winds, but to-day I took leave of
Stuwitz, who was now going to the Banks,
where he intended to anchor some weeks and
make observations on the marine fauna. I took
leave also of my kind friends, Mr. K—— and
Dr. M——; and the latter lending me his boat
and man, I set sail for Long Harbour. The
wind was not very favourable, being south-
south-east, but in hopes it would shift a little
we made a long stretch off to the southward,
and tacked when we thought we could weather
Red Island. It soon however began to rain,
and a breeze sprang up that sent a heavy
tumbling sea after us, and obliged us to keep
away. I went and lay down in the fore cuddy,
a place about the size of a dog-kennel, and
stinking of salt butter and fish, and was
dreadfully sea-sick ; and as soon as we got

between Red Island and Merasheen, being under the lee of the former, we had no wind, and we lay rolling in heavy rain on the tumbling swell that came in after us. At last, by dint of oars, we got to a place called Indian Harbour, and put up at a hut just erected by two old men who had recently come to live there. Their place was clean and comfortable, and they gave me a tolerable bed. One other man and his wife lived in the harbour, which, though safe and convenient, had till lately been studiously avoided. It appears it had the reputation of being troubled, or haunted, which for a long time had prevented its being inhabited. Boats had been known to stay out at sea all night in the roughest weather rather than put in here alone; and no boat ever entered unless in company with others. Although my men pretended to make a joke of it, I could see they were believers in their hearts: their first inquiries were after the spirits, and they were rather disappointed when one of the old men, being an Englishman, said the spirits were all rats.

From the name of the place—Indian Harbour—I am rather inclined to suspect its bad reputation may have arisen from some atrocity either committed upon or by the Red Indians in former days, but could not hear of any tradition to that effect.

June 21st.—Thick fog, but on a little breeze rising we set sail : the fog, however, soon cleared off, the breeze died away, and we lay on a glassy sea under a blazing-hot sun in a perfect calm. We were obliged to take to the oars, which, being small and short, had but little effect on our heavy boat, and we were nearly all day getting across to the harbour of Red Island. This place was composed of red granite. We got comfortable quarters at a Mr. M'Carthy's, and I slept with several of the male branches of the family in a long, low loft, extending the whole length of the house, with a range of narrow beds, or berths, along the wall on one side, and stores and provisions on the other. A little window at one end let in light, and a trap-door or hatchway in the middle of the floor, opening into the kitchen below, let in air.

June 22nd.—Thick fog and quite calm till about nine, when a breeze from the south springing up, we set sail, and had a fine run

past the Ram Islands into Long Harbour.
There were four houses in the middle of this
inlet, but all the men were absent fishing, except one old man.   Getting a guide across to
Trinity Bay was accordingly out of the question, as the old man had never been across:
he said, however, that there was a kind of path
all the way across to Norman's Cove, in
Chapel Arm, and he could show us the com-
mencement of it.   After giving us a dinner of
tea and fish, he accordingly accompanied us
in the boat to the head of the inlet, where there
was a winter-house, and went with me through
the woods to the barrens, where he showed
me a small deer-track, and, the fog clearing off
a little, I got a view of the bearing of the
country for a mile or two round.   Returning
to the winter-house, he went off with the boat
and left us to make the best of it.   Simon and
Tom had made a fire, cut some boughs for
beds, and I set them to get some " rinds" or
bark and repair the roof, and it was fortunate
that we did so, for during the night it rained
very hard.

June 23rd.—The morning was dark, foggy,
and unpromising.   By six o'clock we had

breakfasted and packed up our loads, and, as our provisions were not very plentiful, we determined to advance rather than stay where we were. Climbing up through the wood, we came across the little track and followed it steadily for four or five miles over rocky barrens and mossy marshes till we came to a pond. Beyond this, we had been told of two remarkable hillocks, between which we were to pass. The fog was now so thick that we could not see the hillocks, but I believe we passed between them, as we still had a little track with us, and followed it to a large marsh, when it diverged to the left, and from its bearing I was sure it was wrong. Returning, we found another little track in a better direction, which led us to a small ridge, where it became too faint to follow farther. By this time it was ten o'clock, and the rain was pouring down in torrents, the bushes and marshes as well as our clothes being soaked with wet. We became entangled among several ponds, one of which was of considerable size, in a flat valley ; but, getting upon another ridge, we found a track which led us down to a valley full of woods, in which was a large brook flowing to the north.

In this wood the path ended. Returning to the rising ground, the fog cleared off a little, and I got a sight of Spread Eagle Peak, and found the brook flowed down a well-marked valley straight to the north. This valley I knew must either lead out to Long Cove or Chapel Arm. If to the former, there must be a valley two or three miles farther on leading to Chapel Arm, where I knew that a considerable stream emptied itself. I determined, therefore, to cross the brook and ascend the next ridge to ascertain this point. While spreading about on the side of the valley in search of the clearest space, my men came on a buck, doe, and fawn, of the cariboo, trotting gently along about a hundred yards from them. Unluckily I was on the other side of a small hill, and did not see them. I shot, however, a brace of ptarmigan for supper, lest we should be obliged to sleep in the woods. On ascending the next ridge, we could only see through the mist a large pond with some islands in it, and the ground beyond seemed to rise rather than fall. As I knew, from the aspect of the coast, which I had visited the preceding summer, that any ridge between Long Cove and

Chapel Arm must be narrow, I was now con-
vinced that the valley we had passed was the
right one to pursue.   We accordingly returned,
and kept along the flank of the valley, avoiding as much as we could the thick woods.   At
length, however, we were obliged to enter
them, and slowly, and with great exertion, we
forced our way down to the brook.   Here we
were delighted to find a few strips of long
grass on the margin of the water, and kept
wading across the brook to take advantage of
them as they appeared on either of its sides.
Presently, however, the brook made a sudden
turn to the right, and descended through some
brick-red slate-rocks by several falls into a
narrow and precipitous ravine.   We were then
again driven to the thick woods, and the sight
of a good birch-tree in a small open space almost
inclined me to think of stopping and trying
to bivouac.   Everything, however, was so wet
that we almost despaired of being able to light
a fire, so, taking a mouthful of whiskey we
had brought with us from Merasheen, and a
mouthful of biscuit, we pushed on.   My knee
was by no means well, and began now to get
very weak, and the extreme toil of the woods is

most disheartening under any circumstances.
However, we came before long to a small
lateral valley which led us down to the brook
again ; and its rocky bed, where the water
was not too deep, enabled us to proceed more
rapidly.   It was from ten to twenty yards
wide and about knee-deep, but in another mile
we were again stopped by falls ; a barrier of the
bright red slate opposed its passage, through
which it had worn its way in the most singular
manner.   In some places it was contracted
into a deep foaming channel not above five
feet wide, with a fall of some ten feet, and
then a deep black pool seemed to give the
waters rest for a new leap.   The channel first
made a sudden and rapid turn at right angles
to its general course, the water leaping from
ledge to ledge, and then in about twenty yards
it made an equally abrupt turn back through
walls of the clearest and brightest smooth red
slate.   The descent altogether was not great,
nor was the body of water enough to produce
a powerful effect, but the contrast of colours in
the bright-red slate and the dark-green woods,
the white and yellow lichens and mosses and
the clean dark-brown water from the marshes,

the roar of the waters contrasted with the intense silence of the woods, and the wildness of
the little glen in which we stood, detained me
some minutes in admiration, wet, tired, and
hungry as we were, and uncertain where we
should lay our heads that night. The woods
below this place became thicker and thicker,
until at last we came to a part where it was no
longer possible to force our bodies between the
small pole-like trunks of the trees. This is no
exaggeration, but the simple fact; the trees
would not admit us among them, and we were
obliged to scramble down to the brook and
wade in its waters. We were, however, glad
to find it getting less rapid and wider, and
shortly we came to a still, wide place, where,
on tasting it, I found it brackish, and immediately after we came to a kind of salt-water
pond into which the tide flowed, and going
round it, we stood on the shore of the sea. We
were at the head of an inlet with rocky cliffs,
but at first I could not tell, through the mist,
whether it was Chapel Arm or not. This was
important, as Chapel Arm was the only inhabited place for miles, and that only near its
seaward entrance. Close examination, how-

ever, showed me a headland which I recognised, and we also caught sight of a skiff
at anchor some distance off, just discernible
through the mist, which occasionally cleared
off a little. As it was now quite calm, we
hailed stoutly and fired several shots, and
soon had the satisfaction of seeing the skiff
come towards us. It contained two boys belonging to Norman's Cove, who had been fishing,
but were on the point of going home when
they heard our hail, at which they were much
surprised and at first rather alarmed. We had
hardly got on board before a wind sprang
up from the north-east which would have
prevented them hearing our shot, and before
we had rowed a hundred yards it increased to
a breeze that made rowing useless, and we
were obliged to hoist our little sails and beat
down against it. The inlet is quite open to
the north-east, and we had a jumping sea to
contend with for two hours. When we got on
board, though wet, we were warm with exertion,
but sitting cramped up in the little boat, in a
cold wind, with driving fog and rain, seemed
to numb our very blood. The salt spray, however, beat over us, and perhaps prevented our

catching cold. At length we moored the boat in a little rocky crevice behind a projecting crag, the only shelter she could have with this wind; and by the aid of some posts we climbed up the cliff, being as much as our numbed limbs would enable us to do. Shortly after we arrived at a house belonging to an Englishman named Temple, who received us most hospitably, and gave us hot coffee and dry clothes. It was just six when we arrived at his house, so that we had been travelling twelve hours without stopping. The distance from Long Harbour to Chapel Arm is not more than nine miles, but Temple said we must consider ourselves fortunate under the circumstances in getting out at all, and most fortunate in seeing his boat, as he would have defied us to have got from the head of the Arm to his house (only three miles) before nightfall. A man who had come across in fine weather last winter, when the country is much easier to traverse, slept at the head of the Arm all night, and, though he had the sea-cliff on one side and a steep hill on the other to prevent his straying out of the right direction, he did not get down to Temple's house till

after dinner-time next day. Two gentlemen
with whom I had some acquaintance had been
lost for two days whilst attempting to cross the
summer before, and had been obliged to return
to Long Harbour. Temple told me he believed
they paid the man who at last came to pilot
them as much as 5*l*., although any person who
knew the way could go and return the same
day.

June 24th.—A beautiful morning, and we
went in Temple's boat to Dildo Harbour, where
we found a very fair road down to New Harbour, at which place I put up at the house of
Mr. N——, the gentleman I had seen last year.
The country from Chapel Arm, all the way
through Spread Eagle to New Harbour, is
low and sheltered, and this extreme corner of
Trinity Bay is more pleasant and fertile than
the generality of Newfoundland. A gentleman from Trinity came in, in his cutter, who
said, if we could get horses to take us the first
part of the way to Spaniard's Bay along the
new road, he would go across with me and
return to Trinity Bay by way of Carbonear
and Heart's Content, to which latter place he
sent for his cutter.

June 25th.—We were disappointed in the
horses, the only two belonging to the place
being out in the woods. However, at seven
we set out, Mr. N. and another resident accompanying us the first five miles. A road
twenty feet broad had been cut through the
woods for this distance, leaving the stumps
and the boulders among which a narrow footpath winded. The brooks were bridged over,
however, and it was delightful travelling when
compared with the wild and untrodden woods
and marshes. We then came upon three or
four miles of barrens, across which the road
was not made, a mere track being observable
over their rough and uneven surface. Here
I shot a brace of ptarmigan, which, by the
way, are always called partridges in Newfoundland. We then came into the woods, and
the road gradually improved as we approached
Spaniards' Bay. Unfortunately we found that
Mr. D——, of this place, was in St. John's,
getting his summer supplies, but his house-
keeper gave us some tea and ham, on which
we lunched and set off again for Harbour
Grace. Since I was last at Spaniards' Bay

the new road had been completed, and we
found an excellent road all the way, sufficiently good for a gig to traverse. At Harbour
Grace we were once more in the region of inns
and public accommodation, and the next day
I crossed in the packet to Portugal Cove and
returned to St. John's.

CHAPTER X.

Start for Bonavista Bay—Walk to Harbour Grace, and visit Carbonear — Arrive at Bonavista—Expedition in search of Wood—Clode Sound — Bunyan's Cove and Bread Cove—Greenspond—Advance into the Interior— Mount " Man Point Ridge "—A Haunted Cove—Bloody Bay and Troytown—Ascend the Lonil Hills, and return to Greenspond.

July 16th.—I was detained three weeks in St. John's for want of a mode of conveyance to Bonavista Bay, but was at length offered a passage in a boat which Mr. Packe of Carbonear was about to send to find wood. I accordingly rode over to Portugal Cove, intending to cross to Brigus, and look once more at the geological position of some rocks there, concerning which I thought I might have made some mistake on a first examination, and I afterwards proposed to walk down to Carbonear. At two P.M. we sailed in the Brigus packet, but were becalmed all the afternoon in the bay. Father M‘K——, the Catholic priest of Brigus, was on board, and, as with my usual haste and carelessness I had forgotten to make provision for this lengthy voyage, I should have kept a very Catholic fast had it not been for his kindness. We dined on "fish and vang," which being interpreted means cod-fish and salt pork cut into "junks" and boiled together, and with a mealy potato it is really a most excellent dish. The term fish is restricted in Newfoundland entirely to cod : they ask you whether you will take fish or herring, fish or salmon, and everything but cod and salmon is frequently spoken of by the fishermen as rubbish. We landed in Brigus Harbour about two in the morning, and as there were no inns I became indebted for a comfortable bed to the same kind friend who had provided me with a dinner.

July 17th.—Went down in a punt as far as a small cove called Sculpin Island Cove, where I found my former observations perfectly verified and a mass of the upper or red slate of the country resting quite unconformably on the older grey or St. John's slate. This spot accordingly was a key to much which would otherwise have been very perplexing in the structure of the country. The weather had now been so intensely hot for

some time, and the little strip of fine sand
and the clear green water in this cove looked
so tempting, that I bathed. The water, however, was far too cold to be agreeable, and
one or two plunges quite satisfied me. In the
evening I set off along a very good road,
which was now finished, for Bay Roberts. It
traversed a few wild marshes and barrens at
first, but on coming down to the head of Port
de Grave the scenery was very beautiful.
Patches of grass, clusters of large trees, with
lakes and bold hills, on the one side, and the
broad bay on the other, together with several
small but neat wooden cottages sprinkled about
here and there among the woods, composed a
landscape that reminded me of home, and was
singularly beautiful when seen under a clear
sky, and tinged with the glorious hues of sunset. Arrived at Bay Roberts, I introduced
myself to Mr. Packe's agent there, and was
at once hospitably received.

July 18th. — Walked down to Spaniards'
Bay, and then over the hill to Harbour Grace,
finding an excellent road all the way. There
is now a good road all the way from Carbonear to Brigus, being about sixteen miles.

This is at present the longest and best piece of road in the island; and it is intended to continue it round the head of Conception Bay down to Topsail, in order to open a communication between St. John's and Harbour Grace during the winter. To go from one to the other of these places is at present impossible in winter, except for a strong and active pedestrian, indifferent about fatigue and accommodation.

July 20th.—Visited Harbour Grace Island (on which stands the lighthouse) a second time, and brought away several large slabs of beautiful slate, fine grained, easily cleavable, and splitting into sheets of almost any size. Finer roofing slate could nowhere be procured. While we were there a bit of a breeze sprang up, and we had some difficulty in clambering down the rude ladders over the face of the cliff and dropping into the boat, which was dancing on the waves at their foot.

July 22nd.—Stayed at Carbonear. This place has principally depended on the Labradore fishery, a great number of vessels having been every year despatched from the harbour to that coast. For the last few years, how-

ever, this fishery has almost entirely failed,
or at least the fish have been so partial and
scanty as to entail great loss on those engaged
in it. Several large establishments were shut
up and apparently going to ruin, and the
whole place seemed dull and sluggish. About
a couple of miles north of Carbonear there is
a very pretty valley containing woods and
ponds, and some patches of natural grass.
One day some friends and myself went trout-
fishing in the ponds; and by wading in along
their shallow margins we caught four or five
dozen of trout in a short time. We found very
tolerable paths traversing the woods, and could
not but remark how much easier it would be
to travel in the interior of the island if even
such narrow foot-tracks as these were common.
On the evening of the 22nd, Mr. Biggs (Mr.
Packe's agent) and myself went on board the
boat and set sail.

July 23rd.—Notwithstanding all my sea-
voyaging I was again sea-sick this morning,
and experienced the lassitude and loss of
energy which accompanies it. In the middle
of the day we entered the harbour of Catalina,
where I wished to see what is called in New-

foundland the Catalina stone.   This is nothing
more than iron pyrites, which is found in certain nests and strings in the rock in large
cubical crystals sometimes an inch and a half
wide.   The rock is a grey slate rock, such as
is commonly called greywacke.   As there was
a road from this place to Bonavista, we determined to walk across and send the boat round.
The distance was ten miles, the road broader
than any I had yet seen, but only the two
extremes of it were gravelled over.   Even
the unfinished parts, however, were easy for
foot passengers, the country level, and the
road might easily be made the best in the
island.   All the country around was covered
with gravel, on which grew wet moss, and
the usual stunted, ragged-looking, dense mass
of fir-trees.   Bonavista is a large and straggling but pretty-looking place, with a good
deal of cultivated ground about it, but is sadly
in want of a good harbour.   There is little
shelter even for boats, or this place would soon
be one of the most thriving in the whole
island.   It is much more capable of fertility
and cultivation than the neighbourhood of
St. John's, and nearer to the open sea and

the best fishery than the embayed sites of Harbour Grace and Carbonear. The whole beach, as well as the flakes around, were covered with fish drying in the sun, which the people were now busy piling into round haycock-looking heaps against the approach of night. We inquired for a house of accommodation, and first of all took refuge in a small public-house having only one room, but were shortly fetched away by a gentleman who came in and insisted on our going to Mr. M——'s, where we were most hospitably received and entertained for the night.

July 24th.—We left Bonavista about nine or ten o'clock, but shortly got becalmed off Blackhead Bay. We then got a southerly breeze, and sailed rapidly past the high bold headlands of Keels, which consist of barren rocks stretching far into the interior. The wind then shifted into the south-west, and we beat up towards Barrow Harbour, and in the evening anchored in a small cove under a bare precipice, in one of the Long Islands.

July 25th.—Weighed and stood between the islands for Clode Sound with a light breeze, before which we crept on gradually till

we were opposite Goose Bay, when it became
thick and rainy. Our boat was a large open
one, with a cuddy at each end. The after
cuddy was just large enough to admit a narrow berth or bed-place on each side for Mr.
Biggs and myself, and was about four feet in
height ; the other one was littered with straw
for the men. Each of them opened into a
" standing room" about five or six feet square,
being an open space, the flooring of which
rested on the ballast. The " midship-room,"
or hold of the boat, was covered with loose
plank, and contained our stores. When it
rained we were obliged either to expose ourselves to the weather, or screw ourselves into
the cuddy, while it was absolutely necessary
that we should take all our meals " al fresco,"
for want of room in the sheltered part. The
rain, however, this afternoon cleared off, and a
smart breeze sprang up from the south-west,
which soon became so fresh, that, after beating
against it for some time, we ran into Brown's
Cove and anchored. Here Mr. Biggs and I
took our guns, went up a small brook and
shot a black duck or two : we then traced the
brook up to a large pond, and returned through

the woods by an old wood-path, completely
drenched by the wet bushes.   Meanwhile the
men had been catching lobsters, and had taken
a punt full, by means of a pole with several
hooks at the end of it.

July 26th.—A most beautiful day, blazing
hot, with light variable winds. We sailed all
day up the calm waters of this beautiful inlet,
and found its shores varied with rocky precipices and low banks covered with dense wood.
We landed at several places in a small flat-bottomed boat we had brought with us, and re-
joined the vessel as she sailed slowly on.   On
getting at the head of Clode Sound into a wide
expanse, we sailed to the south arm and gently
grounded in the mud, about a mile and a half
from land.   As the tide was falling, we had
some difficulty in towing the vessel off again,
when we anchored in about ten feet of water.
Going ashore at the mouth of a small brook
on the western side of the arm, I was surprised to find a number of flowers of a most
beautiful kind of convolvulus trailing on the
ground.   The flower is as large as the convolvulus major of English gardens, the colour
white, fading into streaks of delicate pink and

flesh-colour towards the base and margin of
the cup. About sunset, as I sat on a rock
waiting for a shot at some ducks, the mosquitoes began to swarm about me in considerable numbers, when several dragon-flies,
apparently identical with the common libel-
lula of England, came to my rescue. They
flitted about, hawking' after the mosquitoes,
dashing occasionally close to my face, and
catching one here and another there, and as
soon as the prey was pounced upon, they
hovered and balanced on their wings till they
had eaten him all up, body, legs, and wings,
in a manner most delightful to behold. I remained quite motionless, compounding for a
little blood-sucking, in order to see the rascals
thus punished, and thought seriously of taking
a small body-guard of dragon-flies into my pay
immediately, had I but known how to intro-
duce proper discipline and prevent desertion.

July 27th to 31st.—The object of the men
in coming up to the head of this deep inlet
was to look for trees of sufficient size to form
the beams of a brig Mr. Packe was building.
Although there are no permanent inhabitants
nearer than Barrow's Harbour many people

come up in the winter to reside, either for the
sake of fire-wood or to cut timber. We found,
accordingly, many wood-paths and some old
houses and huts in the woods near the shore,
and all the best and largest timber was there
cut out. The beams the men wanted were to
be ten inches square, and they were obliged to
range the woods for many miles alongshore
before they could find trees of sufficient size.
In the mean while Mr. Biggs and I, with
Simon and Tom, traversed the shores of the
sound as far as we conveniently could in the
small flat-bottomed boat, the timber-men taking
the punt. The shoal mud-bank at the head
of the arm was traversed by a narrow winding channel of deeper water, through which
flowed the river that came out of a valley
to the southward. Up this we proceeded ;
but on reaching the fresh water found only a
broad channel of boulders, with the shallow
wide brook flowing and fretting among them.
We travelled over the boulders for some
miles, shut in on each side by dense forests,
which contained many trees of good size,
birch, spruce, and juniper, large enough for
the beams we wanted, but which it would

have been impossible to drag out along the rocky brook. There were several broods of shell-birds, of which we shot some; and at our farthest point we disturbed a fine "gripe," or eagle, who had just killed, and was eating, a shelldrake. He flew up to a cliff out of our reach, where I examined him with my pocket-glass. His feathers were brown rather mottled with white, his head was light-coloured and rather bald-looking, and his strong legs and talons were yellow. Wishing to examine him more closely, I put a pistol-ball into my double-barrelled gun; but, as he was a hundred yards distant, I did not succeed in hitting him, and he slowly wheeled off into the air, and disappeared over the woods. The next day we examined the bight called the northwest arm, the head of which was likewise shoal, and two brooks, a large and a small one, flowed into it. We saw a punt go off down the sound as we left the vessel, and on getting to the brook found against an upright bit of rock a roof of birch bark, which served as a temporary shelter to salmon-fishers. In crossing the arm we passed through an immense shoal of medusæ; the water was beautifully

clear, and was entirely filled for a great depth
by these cup-like animals, slowly·flapping their
gelatinous bodies, and proceeding all in the
same direction. The shoal was half a mile
wide, and there must have been some millions
within that space. Both the brooks flowed
over ledges of sharp slate rock, that made
walking almost impossible ; they were likewise full of rapids and waterfalls, and the
wood about them was of the densest and scrub-
biest character. On returning we landed on
a large flat tract of ground where the wood
had been burnt. Most of the trees remained
standing, but were dry, white, and brittle, the
charred surface having fallen off from time,
and decomposed. The soil was sandy, and
would, no doubt, bear excellent potatoes.

One of the men having reported a large
pond on the south-west side of the inlet as
visible from a ridge he had been on, we went
in search of it. A mile or two of wood-path
led us on to some marshes, and then into
some dense tangled thicket with no path. It
was tremendously hot, and the mosquitoes
swarmed. By smearing our hands, faces, ears,

and neck with tar and sweet oil, which we had
mixed in a bottle for the purpose, we in some
degree avoided them, as they dislike anything
greasy or strong smelling; but we were obliged
to stop every half-hour to put on a fresh dress-
ing. On the marshes we saw some very pretty
flowers, some yellow and some white, appearing to my unbotanical eyes something like
the orchis. There are several other kinds of
flowers on the marshes; and the calmia, the
azalea, and the beautiful wild dog-rose grow
in the woods in the height of summer. We
reached at length a small brook, up whose
rocky ledges we proceeded till we came to a
pond about half a mile long that effectually
stopped our progress, as its margin was too
deep to walk along; and the bare idea of
walking through the bushes that surrounded
and overhung it was enough to sicken any of
us. We caught a few trout, and then proceeded down the rocky ledges of the brook,
which was full of small waterfalls, deep pools,
and rapids. The rock was principally a bright-
red slate rock. We were thoroughly soaked
by two or three thunder-showers, which also

made the rocks so slippery, that in one place, in scrambling over some fallen blocks against the side of a cliff, I slipped between two that leaned together; and, had I not caught by my arms, should have fallen down into a kind of dungeon some eight or ten feet deep, whence my companions would have been rather puzzled to extricate me. On arriving at the mouth of the brook we put off into deep water, where I bathed : but one plunge was quite sufficient, as the temperature of the water, even in this sheltered and shallow inlet, was only 46°, while the air had never been below 70° during the day for at least a month.

The next day we again visited the south-west brook, and found a number of very tidy winter-houses. I then scrambled through the thick wood that surrounded them into some rather more open wood, at a little distance from the shore ; and we made our way to the top of a little rocky hill on the east side of the brook, which was the only bare place to be seen anywhere about. From this eminence we had a fine view. To the east of us were many hilly ridges and broken and rocky high lands, being those at the back of Trinity Harbour,

and running thence to Keels. In the south-
east some fine hills were seen, probably those
around the head of Random Sound; and in
the south was a range of distant hills, either
those between Bay of Bull's Arm and Piper's
Hole, or between that place and Fortune Bay.
To the west and north we had a wide range of
country, in which there were no hills visible,
the land consisting of low undulating ridges,
running about north-north-east and south-
south-west. The whole of this tract seemed
covered with a dense mass of dark wood, with
lighter tints of marsh-land peeping out here
and there. The hill we stood on was probably
not more than three hundred feet above the
sea; so that our view to the westward, from
the nature of the ground, would not reach
to a greater distance than about fifteen miles.
On returning to the mouth of the brook, by
choosing a spot where the fresh-water coming
in raised the temperature of the salt-water
sufficiently, we got a very pleasant bath.
Numberless salmon were leaping here, jumping completely out of the water a yard into
the air, and we stayed some time hoping to
get a shot at one, but none rose sufficiently

near us. In the evening, when we came on
board, we found that the men had succeeded
in finding and cutting ten trees of the required
size, but none of them within a mile of the
landmark: they had accordingly had great
labour in dragging them out.

August 1st.—Sent all hands to bring down
the timber, which we hoisted on deck, and then
got in fire-wood, water, and more ballast, to
balance the deck-load. At noon we set sail
down the sound, but, seeing some fine wood in
Bunyan's Cove, went in and anchored there.

August 2nd.—Mr. Biggs and I found this
morning a brook nearly dry, but containing
deep holes which were full of trout. We caught
a few with bait, but, as the rest were frightened
at this and declined to bite any longer, we determined on "jigging" them. Shortening our
lines, we drove the trout into a corner of a hole,
and then gently dropped our hooks among
them, and, jerking smartly, caught and drew
them out. In this way, in about three hours,
we had got ten dozen, which, as we had no
fresh provisions left, were very acceptable. The
rocks were partly a chloritic slate and partly a
coarse dark-red sandstone. By the aid of fre-

quent applications of tar and oil we had managed to defend ourselves from the mosquitoes
during the day, but at night when asleep in the
little cuddy, the door of which could not be
closed without risk of suffocation, the gallinippers worked their revenge on us. Not an un-
interrupted half hour of sleep could we get:
their dreadful hum, more sleep-dispelling than
the roar of a lion, and their stinging bites, with
the burning irritation of the old lumps and
wounds, in the hot, close cabin, almost drove
us mad. If we went outside and lay on the
sails, they were worse; we had brought no sheets
with us, so we were compelled to roll ourselves up in blankets, hermetically sealing
every aperture, and, bagging it out round our
faces, breathe through it as well as we could.
If, in tossing about during the night, the least
corner of the blanket got loose, they came
streaming in by two and threes, and fastened
on every inch of skin they could find. This had
been our tormented condition for the last week,
and I now felt quite ill and fevered, so much so
that on August 3rd I scarcely quitted my berth
all day, trying to gain some continued rest and
sleep. The vessel dropped down the sound to

another cove in search of a tree or two to re-
place some of those they had cut, which scarcely
came up to the required size.

August 4th.—Just before sailing this morning we landed, and, coming to a little brook, I
fastened a hook to a piece of twine I had in my
pocket, and, tying that to a small stick, we
jigged two dozen of trout for our breakfast.
We then dropped down to Bread Cove, where
we landed, and went into a pond about half a
mile in search of game, but were unsuccessful.
At the mouth of a little brook in this cove
there were signs of former habitations, a cleared
space or two, namely, in which raspberry-bushes
were growing, and in one spot we found a grave
neatly railed in and covered with wild roses.
A piece of plank had been raised for a grave-
stone, on which were two initials carved, and the
date of 1755. Sailing past the inlet of Goose
Bay, we were much struck with the beauty of
the scene, the tranquil waters and thickly-wooded shores of the two inlets, and the varied and
picturesque groups of hills which appeared in
the distance at the head of Goose Bay. The
physical features of the view all round were
beautiful and highly picturesque, but how dif-

ferent would our feelings have been towards it
could we have pictured to ourselves towns and
villages, fertile fields and happy homes, hid in
the recesses of the hills and scattered along the
shores of the sea ! The known barrenness, rug-
gedness and wildness of the country were now
constantly associated in our minds with its varied outline and even its seeming richness and
verdure. We ran down to Barrow Harbour,
the land about which is very bold and lofty,
and from which, consequently, the wind came
off in sudden squalls and gusts. We were then
going to stand outside the islands to run down
to Greenspond, as no one on board had ever
been through the channels of the islands. As,
however, I had Bullock's large chart with me,
I offered to pilot them through, so we hauled
our wind, and, keeping close by Salvage, made
for the entrance of Bloody Bay. When off
Damnable Bay, however, the wind freshened
and veered a little, obliging us to beat up, and,
as night was now coming on, it was requisite
to look out for a harbour. The harbour at the
north end of Morris's Island seemed the most
convenient on the chart, so I told them to bear
away, run past a certain point, and, avoiding

a shoal on the starboard hand, to anchor in
the small cove they would find there in so
many fathoms water.   On all my prophecies
coming true as to the nature of the place, the
men were in great wonderment, as they have
no idea of sailing by chart, and I found it very
hard to persuade them that I had never been
there in my life before.

August 5th.—We sailed down this morning
among a multitude of small rocky islands with
woody heights, through narrow and intricate
channels, in smooth water; and by the aid of the
chart, which was minutely accurate, found our
way into Bloody Reach, when we had a straight
course down by a broad channel called the
" Cow-path " through the islands to Greenspond.   Sailing among these islands in fine
weather and with a fresh south-west breeze is
delightful work;   constant change of scene,
from the opening and shutting of different channels, and the shifting of the woody peaks and
rocks around, and the attention required to steer
clear of changes and find the proper way among
them, excites the interest at every turn.   On
getting outside of them, although there had
been so long a continuance of hot and sunny

weather, we saw a large iceberg aground near the Gooseberry Islands.   At eleven o'clock we landed in Greenspond.   This is a straggling place on an island of granite, the little harbour being formed by several other smaller islands. There were several good houses and large stores, with a very decent church, and with planters or fishermen's houses, neat, clean, and larger than usual, perched here and there upon the rocks or dropped in the hollows.   A tolerable road had been constructed through great part of the place, but the inhabitants were in great want of fresh water, having to fetch much of their supply from the mainland, three miles off.   There were several brigs, brigantines, and schooners, all busily loading with fish, and one large brig unloading a cargo of salt.   Mr. Biggs was to leave me here and go on to Fogo, if I could hire a boat, which, after some little difficulty, I succeeded in doing.   Meanwhile, Mr. W., (agent for Robinson, Brooking, and Garland,) having heard of my arrival, sought us out, and in the kindest possible manner insisted on my making his house my home during my stay.   Dr. W. and the other residents likewise, were very kind and hospitable.   They were all English in this harbour, and all members of the Church of England, and on the Sunday the church was very well filled. The clergyman of the place, however, had been recently removed to Bonavista, and a licensed reader took the duty till the appointment of another or till he himself was ordained.

August 6th to 12th.—After having got ready the boat for a start, and hired a steady old fellow, named Robert Saunders, as skipper and pilot, I only wanted a short start of fair wind to run up the bay and visit some other of the islands and inlets of which it is full. A steady southwest wind, however, blew the whole week, and one day it rained very heavily. Had I had a vessel of my own this would have been the best possible wind for me, as I could have worked regularly and gradually alongshore, and should not have been obliged to overrun my work by coming on to Greenspond, where, having once arrived, I was detained during the whole week.

Mr. W. told me he had formerly had a tame deer which had been caught when young, and been easily and perfectly domesticated. It wandered about the island, which is about four

miles across, but on his going on to a rock and
hallooing its name, " Tallyho," it would jump
out of the thicket it happened to be in, climb
on to a rock, and, as soon as it saw him, come
galloping and bounding up to be coaxed and
fed.   Two or three times, however, it strayed
away, and swam across to the mainland or the
neighbouring islands, and at last it went away
on the ice and was seen no more.   Had he
had several of them to form a herd, I think
they would have become completely domesticated, and, in that case, if they had gone off,
they would probably have returned.   It would
be a very interesting experiment to get some
on a lonely island, in which there were no
dogs, and tame them; and as they are nearly,
if not quite, identical in species with the reindeer of Lapland, they would become highly
useful, and, I think, be a means of turning the
interior of Newfoundland to profit.

August 12th.—At last a light air sprang up
at five o'clock this morning from the north, by
the assistance of which we got out of the harbour, sailing past Shoe Cove Point and the Fair
Islands into Locker's Reach.   Here I took the
punt and went round the Frying-pan Island to

get a shot at some puffins that were flying
about. The rocks were all composed of coarse,
largely crystalline granite, red inside, but weathering white; and in the passage between
Trinity Island, or Lewis's Island, and the
Frying-pan, the bottom of the sea consisted
of huge peaks and mounds of this white granite, rising from dark and deep hollows.
The extreme clearness of the water rendered
these cliffs - and peaks all visible as we approached them, though none reached to within
three or four fathoms of the surface, and the
sensation experienced in sailing over them
was most singular, and to me very uncomfortable. I could not look over the boat
without extreme giddiness, as if suspended on
some aërial height leaning over a tremendous
gulf. The same sensation was described to me
by a gentleman I afterwards met with, an experienced hunter and sailor, as assailing him
upon his once in smooth water taking a boat
within the space of some sunken rocks off the
Whadham Islands, on which the water broke in
bad weather. These rocks he described as
three peaks rising from an apparently unfa-
thomable depth, and the sensation, as his boat

gently rose and fell between them, was so un-
pleasant, and indeed awful, that he gladly got
away as fast as he could.   From Locker's Reach
we ran into Content Reach, and thence through
a narrow opening into Freshwater Bay, up which
we sailed with a fair breeze and a bright sky, and
anchored on the bar of a large brook that runs
in at its head.   At the mouth of this brook was
a great salmon fishery, and there were several
houses inhabited by men attending to the nets,
and the coopers who made the casks for packing the fish.   Our boat was brought by a
winding channel to the mouth of the river,
and moored to the head of a small stage.
There was much rank natural grass growing
about the mouth of this brook, and two or
three boats were there whose crews had come
to cut it and carry it away.   Some twenty or
thirty years ago there had been a large brick
house here, the inhabitants of which owned
the salmon fishery and had a considerable
cattle-farm ; but they had decayed and left the
place.   I slept on a bench in one of the fishermen's huts, four other men lying on the floor
before the fire.

August 13th.—As the salmon fishery was

now nearly finished, George Lane, one of the
coopers, agreed to pilot us up the brook to
some large ponds; so, borrowing a good punt
with sails and oars, our own being rather
rickety, we set off at six this morning. The
brook flows down a valley about two miles
broad, on each side of which is an abrupt
ascent of about 150 feet covered with wood.
The bed of the brook is wide and full of
boulders, and divided into several channels,
and at this season of the year it is very low.
Proceeding with the oars for two or three hundred yards, we came to the ledge of the first
rapids; we had then to get out on the boulders, drag the boat, and push with poles into
a narrow foaming channel and up a "shute,"
or small fall of a foot or two: then came a
succession of small pools or steadies, shallows,
rapids, and shutes, for about two miles, the
river altering its direction twice in that distance. These two miles cost us two hours'
hard labour, having frequently to get out and
almost lift the boat up some of the shutes. Nothing but George Lane's intimate acquaintance
with the winding channels and those passages
in which there was most water would have
enabled us to get up. On arriving at the head

of the brook we came to a fine pond or lake
about a mile broad and nine miles long, turning
short on our right hand, and running straight
into the country at nearly a right angle with
the course of the brook.   In another mile we
came to the mouth of a little brook, running
into the lake on the right hand, and called Mint
Brook, from the abundance of a kind of mint
growing on its bank.   We hoped here to find
geese or other game, but saw only a large
family of otters in a rocky pond, and these,
when once disturbed, took care to keep out
of shot.   Above this little pond the brook
formed several very pretty waterfalls.   We
then proceeded up the pond ; it was bounded
by woody cliffs rising from one to two hundred feet, the tops of which were on the general
level of the country.   Near its head we saw
two old geese and five young ones, and gave
chace, but could only come up with one young
one, which I shot.   When nearly approached,
the geese "bill" as the men call it, that is,
stretching out their neck, they sink themselves
beneath the water, swimming away with the
bill only just above the surface, and, when
fired at from a boat in that position, the shots
glance from the water frequently without

touching them.   The only way then is to
stand up in the boat and fire just under the
bill, so as to hit them in the head.   On arriving at the head of this pond we found another
brook running into it, of much the same character as the one below, but not more than
half a mile long.   This comes out of a steady
channel that opens into another large pond
similar to the last.   As we were hauling the
boat up the last shute of this brook, Simon
and Tom slipped and let go their hold, thus
letting her head sweep round, and exposing her
broad-side to the rapid, which tore her out
of our hands, and away she went broadside
down, and would have capsize dover the rocks,
and been swept away, had not Lane rushed
into the water and forced her stern round,
when he was able to hold her till we came to
his assistance.   We should have been in a very
pretty " fix, " as the Americans say, had our
boat and provisions taken their departure.   On
getting to the upper pond the wind died away
and left us fully exposed to the blazing sun
above, and its reflection in the glassy water
below.   The heat was so great that the men
could scarcely row, and my face was blistered

and the skin peeled off my lips. However, about sunset we arrived at the head of the pond. Here there are two valleys, and the brooks running down them were now shallow; but Lane said that in spring, when the waters are high, a flat-bottomed punt can go fifteen miles up the south-west valley. We landed on a strip of sand-beach, where Lane said there was a winter-house, but we found it had fallen in from age; we cleared out the rubbish, however, and made a fire on the hearth, where we cooked our supper and slept " *sub Dio*." These ponds are called " Gambo ponds," and their banks are generally inhabited during the winter by several families, who bring in their supplies and cut timber, or shoot the deer as they cross the pond in spring and autumn. We found one or two platforms about the ponds, raised ten or twelve feet high on four posts and fenced in with boughs on which the men stand to watch for the deer. As soon as from these look-outs they see a herd coming out of the woods and taking to the water, they pursue them in boats, and generally kill three or four at least.

August 14th.—At six we set out up the

north-west brook, which admitted our boat for
about half a mile, when we left it and struck
into the country.   Through a little skirt of
thick wood we reached a more open part,
where the wood had been burnt many years
ago. Some very fine stumps of trees were
still standing, showing that good timber will
grow in this part of the country, and the
young trees were principally birch.   In about
a mile we reached the top of a bare hill, but
not having a good view we proceeded along a
fine deer-path to a higher ridge, from which
we could see all that was to be seen of the surrounding country.   Its character on a nearer
view was the same as it appeared from the hill
in Clode Sound.   It consisted of long undulating ridges rising two or three hundred feet
high, their slopes covered with wood, their tops
often forming barrens, and the intervening valleys generally occupied by marshes. No distant
hills were visible, and George Lane said that
the old furriers used to have a tilt on one of
the brooks forty miles inland, reckoning from
the salt water.   We were now twenty miles at
least in the interior, and there appeared no in-
ducement to proceed : we accordingly returned

by the deer-path, and soon got down to our boat.
A deer-path is like a sheep-walk on a common
—a narrow winding track about six inches
wide. The one we traversed must have been
made by a very large herd, as it was hard and
bare even on the marsh; and a herd of some
hundreds had been seen to pass along it the
preceding winter.

Returning down the pond we had a pleasant breeze that carried us quickly along. We
landed at one place to examine a large land-
slip on the face of a cliff; at the foot of which
piles of rubbish, stones, and trees lay heaped
in confusion. All the rocks on the borders of
these ponds were varieties of chloritic and fine
micaceous slates. Under the skilful guidance
of Lane we shot the rapids, sailed down the
second pond, and then shot down the lower
brook without meeting with any accident.

August 15th.—Being obliged to wait for
the tide to get over the bar, we did not sail
till the middle of the day, and then had
a very strong west wind. We anchored
under the headland between Freshwater Bay
and Cat Bay, and climbed up Man Point
Ridge. I was now become pretty well accus-

tomed to the country, and could lead most
of the men with whom I went into the woods,
but of all the scrambles I ever had this of
Man Point Ridge was the worst.   It was
not more than 500 or 600 feet high, and not
steep, consisting of a succession of receding
rocky ledges, like gigantic steps.*   The little
cliffs were easy enough, but the flat places and
slopes between them were almost impenetrable.
The old wood had all been burnt, and the
sharp dry stakes and trees and branches had
fallen across each other in all possible angles
and positions, making of themselves a perfect
stockade.   Among them, however, was growing up a new and still more dense vegetation
of fir, birch, alder, &c. &c.   Climbing over
the bushes from one white slippery stump to
another, crawling beneath them between the
roots of the young trees for ten yards at a
time, wading, pushing, and tearing through all
when we could neither crawl nor climb, with
the thermometer at 80°, and clouds of mosquitoes obliging us to stop every now and then for
a fresh smear of tar and oil—such was our progress for an hour and a half, when I contrived

* The rock was a fine grey slate rock or greywacke.

to reach the top, the distance from the sea-shore being in a straight line about a mile. Simon and Tom did not get up for half an hour longer—indeed, I met them as I was beginning to descend. The view from the top was very fine over the surrounding islands and bays, but not equal to what I had expected, and no objects of interest were visible towards the interior of the country. In going down we came to a small marsh, where I shot three ptarmigan; and we got back to the boat in about four hours from the time we started. The wind had now moderated, and we sailed round Man Point and anchored in Dog Cove. Here we found a very good winter-house about a quarter of a mile in the woods, with abundance of excellent raspberries, but more mosquitoes; and as we found we could not eat without being eaten, we retreated to our boat. Our boat was like that previously described belonging to Mr. Packe, but smaller, and the two end cuddies were only large enough to stow provisions in. I had no bed, but only a blanket with me: we accordingly cut some boughs, and strewed them in the hold, and slept upon them. The covering of the hold was merely a number of loose planks, but we had brought a large piece of tarpauling with us to stretch over it if necessary. This had however been rolled up, and, having been trodden on occasionally, was now, in consequence of the extreme heat, fastened and glued together. During the middle of the night we were awakened by a tremendous thunder-storm, and the rain poured down in bucketsful: we jumped up, but could not spread open the tarpauling in the dark; and we accordingly sheltered ourselves the best way we could under a corner of it, and made the most of our bed of boughs, which were soon wet enough. My blanket sheltered me a little.

August 16th.—A beautiful morning, and, taking advantage of it, we dried ourselves in the sun, and hauled off the planks to admit his rays to our bough bed and sleeping apartment, from which there was shortly rising a famous steam. We landed to fill our water-casks, and I followed a narrow path into the woods for some distance, but got nothing for my pains but another thorough ducking from the wet bushes. We then sailed to " the Beaches," a small cove at the mouth of Bloody Reach, where, as the wind and tide were both against us, we anchored till the latter should turn. We got plenty of currants and raspberries here, and had a delicious bath, this being the first time I had found the pure sea-water warm enough to be pleasant. At the turn of the tide we sailed, and beat up Bloody Reach till dark, when we anchored in a small cove on the starboard hand, called Goose Cove. Robert Saunders, our pilot, a stout old weather-beaten English fisherman, had a great reluctance to go far into this cove, and gave as a reason the number of flies there, and we had hardly dropped anchor inside the headland before these gentry trooped off to us in myriads. He at last gravely assured me that the place was haunted: many people who had gone to the head of the cove and moored to a rock or a tree had had their ropes cast loose again as soon as their backs were turned, and this many times successively. Others had been disturbed by nocturnal noises and shrieks resounding through the still woods. He himself when a boy was with his father here one night, when, after being frequently disturbed by shrieks and cries, they put to sea again, unable to endure it longer.

Fortunately the ghosts and ghostesses did not think proper to pay us a visit, and we slept soundly till the dawn.

August 17th.—Before the sun showed his honest face we had worked up two or three miles, hoping to reach the brook at the middle arm of Bloody Bay with the morning tide, which we accomplished. The land round Bloody Bay is steep and lofty, covered with a thick vegetation of young trees, while bays and arms of water run among the hills in every direction. The country had all been burnt twenty or thirty years ago, and the ghosts and skeletons of the old trees rose white and spectre-like among groves of light-green birch or more sombre-hued fir, spruce, or pine trees.

On arriving at Bloody Bay main brook we found a very decent man named Stroud, with his wife and seven daughters, the oldest not more that twelve years. He was the only summer resident in the bay, except one old man, a cooper, who made casks for the salmon. Stroud attended to the salmon-fishery, which belonged to Messrs. Brooking and Garland, from whom he received regular wages, and a dollar for every tierce of salmon he caught.   He had this summer caught forty-six tierce besides those consumed by his family. This was reckoned a very great catch for the mouth of one river.   He had a comfortable house, a few cattle, and several very pretty little cleared spots or gardens, in which grew abundance of excellent potatoes, cabbages, greens, and turnips.   The flat land on each side of the brook is half a mile wide, and is of good quality.   Deer and game of all sorts are very abundant at the proper seasons; and he said he generally made 20*l.* during the winter by the sale of game and furs.   This, with perhaps 30*l.* for his summer's work, his house and land rent free, and all his provisions raised by himself, except bread, a little pork, tea, sugar, and molasses, were certainly enough to put him in a condition which many families in England would be but too happy to realize for themselves; but then it must be remembered that none except those born and bred in the island would be able to make anything of it.   There was some very fine timber up the valley—one birch-tree at four feet from the ground measured seven feet in circumference: this, however, was the largest I saw, and the large trees only grow singly here and there among the usual stunted undergrowth. The brook is from eighty to one hundred yards wide, but much encumbered with rapids. We walked through the woods and marshes for some miles along its banks, and I regretted I had no canoe or other suitable boat to explore it farther. There is a large pond some fifteen or twenty miles in, which Stroud called Terra Nova Pond, and which he said was twenty miles long.* No place struck me as so suitable for an exploratory expedition into the interior of Newfoundland as this. By sending up a large store of provisions to the head of Terra Nova Pond (if it answer Stroud's description), and making this the head-quarters, and getting a good Indian as a guide, excursions might be made out to Bay Despair on the one side, Gander Bay on the other, and probably to the large lake, called by Mr. Cormack Jameson's Lake, in the centre of the island. Were I ever again to visit Newfoundland, this would be the point I should select next after the head of White Bay; but it would require a

* Further information on the geography of the district is given in the Report at the end of the section of Physical Geography.

F 3

good party and preparations, and, above all, a Micmac Indian or two. With a good Indian hunter, even a sporting party in the months of September and October would I think be amply repaid for their trouble. Stroud was greatly surprised that we saw no deer even in the little distance we went, as he showed me several spots where he had killed them at different times. He had killed likewise several wolves lately, and not more than a week before our visit his children came running in one day to their mother quite frightened, and said a large strange dog had come out of the wood while they were playing on the beach, and went up towards the calf which was feeding close by, but on seeing them turned, and growled and snarled at them, till on their running and screaming he retreated again into the wood. When Stroud came home at night he took his gun, and found at the place mentioned the track of a large wolf, which he followed some distance into the country. He had now set a trap for him, baited with seal's flesh, on a marsh close by.

August 18th.—After breakfast I sent Simon and Saunders round in the boat, and took Tom

with me across a narrow neck of land to the
south west arm of Bloody Bay, which is called
Troy-town. There are no permanent inhabitants in this place, and though several families
generally winter there, there is nothing like
a town, and I did not learn why it was
called Troy. The neck of land between the
two arms was about a mile and a half across,
with thin skirts of wood round a marsh, leading on to some barrens. Here I put up
three coveys of ptarmigan, and bagged three
brace out of them. A little brook flowed out
of the marsh to the edge of a perpendicular
cliff about eighty feet high, over which it flung
its waters in a beautiful cascade. On either
hand the cliff gradually rose to double or treble
this height, and the high grounds at the back
were covered with wood. At the foot of this
great wall, and below the cascade, there extended a flat district, beautifully dotted about
with clusters of birch; and beyond, in the
hollow of bold and picturesque hills, lay the
bright waters of Troy-town, calm and unruffled
as a lake. On the left hand of this was the
narrow entrance into the waters of Bloody
Reach, and on the right the lofty eminences

of the Lonil Hills, sweeping in bold ridges
down to the margin of the inlet, and casting a
broad shadow over half its space. I stood on
the edge of the cliff, struck with the singular
and picturesque beauty of the scene, and utterly
regardless of the " conk" of half a dozen geese
in the waters below me, who, even at this distance, were alarmed at my figure, and lazily
took to flight. On getting down to the strip
of flat land which here stretched along the foot
of the precipice, we found it covered with most
excellent whortleberries, hanging in clusters
on bushes about a foot high : they were larger
than the largest black currants, and of a rich
juicy flavour. As it had fallen perfectly calm,
and there were no signs of the boat, we dined
upon these berries; but having unfortunately
forgotten the tar and oil, we. afforded a rich
repast to the mosquitoes, until we contrived
to light a fire on a point of rock jutting
out into the water, when the smoke relieved
us in some measure from their attacks. In
the afternoon Stroud and his boy rowed
round in a skiff, and came down to us, when
we cooked a ptarmigan, and had some tea,
and in the evening our own boat made her

appearance off the inlet, and anchored in a small cove near its entrance. The depth of water at the entrance of Troytown is only sufficient for a small skiff, and much of the water inside is very shoal. Stroud was come round to cut grass, which grew in a narrow band just along the flat shore; so, borrowing his punt, we went off to our boat. In the cliff above the small cove where we were anchored a fish-hawk had her nest, and kept slowly wheeling round, uttering every now and then a monotonous and dismal screech. This was a sufficient explanation of the noises in the wood at Goose Cove, but Robert Saunders shook his head, and, though he said nothing, I could see he by no means admitted the force of it. I need scarcely add that there was a strong muster of mosquitoes on board to-night.

August 19th.—We went up Troy-town again this morning, and after examining its shores we landed to ascend the Lonil Hills. This was a desperate bit of work: we tried at one or two points before we could succeed, but at last struck in by a little brook that came out in a sandy cove. The sun was burning hot; there

was no wind; the woods were either so thick
as to be scarcely penetrable, or, where they were
thinner, a green close shrub with a small leaf
and stiff branches formed a brushwood breast-
high.   On arriving at the foot of the hill we
found its sides very steep, obliging us often
to use our hands in climbing; but about half
an hour placed us on the top.   The view was
very beautiful; and I should estimate the hill
at about 800 feet above the sea.   We could see
the entrance of Clode Sound, Newman's Sound,
and a great number of hills in the distance beyond them; all the islands filling Bonavista
Bay to the north of us, a great extent of undulating ground towards the west, besides the
lovely home-view of Troy-town immediately at
our feet.   A distant range of hills bore true
south-south-west, and a very distant peak-like
centre hill at the head of Trinity Bay bore
exactly 206° 30′ by prismatic compass, or as
nearly as possible true south of us.   On reaching the beach again we had a pleasant bath,
and then rowed down to our boat, took in wood
and water, hoisted the anchor, and away we
went down the bay before a pleasant breeze.
On reaching " the Beaches " we hauled round

into Content Harbour, where we anchored.
The view along Bloody Reach is very beautiful : it is a straight channel twenty miles
long, and a mile or two broad, through numberless islands of all shapes and sizes, with
one little conical island, called Mouse Island,
just in its centre. A glance at the map will
show something of the singularly indented and
complicated outline of the shores of Bonavista
Bay, with its numberless islands, inlets, bays,
headlands, coves, and rocks; but Bullock's
large chart is requisite to give a good idea of
them—a chart, the construction of which must
have been a work of immense labour, and
which is singularly and minutely accurate.

August 20th.—Very unwell last night and
this morning, with sickness, stiff neck, and
headache. However, we landed in Locker's
Bay at Chalky Cove, to examine the cliffs.
This place is so named from its white cliffs,
which consist of granite, weathering white
outside. The whole of Bonavista Bay, from
Locker's Bay to Cape Freels, and beyond, is
composed of excellent granite, which might be
worked in any quantity and in blocks of any size,
and would make a very handsome building-

stone. (See the Report.)   From Chalky Cove
we sailed through Trinity Gut, inside Lewis's
Island, and then ran through the Fair Islands
down to Greenspond, which we reached about
five in the afternoon.   On the Gooseberry
Islands coming in sight, I looked for the large
ice island which I had seen off them, and which
had remained quite stationary and apparently
unaltered during my stay at Greenspond.   It
was, however, gone, but how, when, or where,
nobody could inform me.

To give a general idea of the value of money
and price of labour in this part of Newfoundland, I may as well mention here the expense
of this trip. I gave the owner of the boat 7s. 6d.
a-day for her hire, and Robert Saunders 6s.
a-day and provisions for his services, altogether amounting to 4l. 17s. 6d. currency
(rather more than 4l. sterling) per week for
the mere hire of an open boat and one man.
The first demand was more than this; but
then it was the only boat and almost the only
man in the place that could have gone, and I
considered myself lucky in getting one at all.
A month earlier I could have got neither, had
I offered double or treble the sum.

CHAPTER XI.

Expedition to Fogo Island, Toulinguet, and Exploits River
—The Red Indians and the Mic-Macs—Sporting Excur-
sion—The Bishop's Falls—Buchan's Island — Return
without reaching Red Indian Lake—Arrive at Toulinguet
—Description of the Court-House—Start for the Bay of
Exploits—Reach Fogo Harbour and sail for St. John's—
Return to England.

AUGUST 24th.—As there were no vessels or
boats in Greenspond going north, I was obliged
again to hire the boat, with a crew of four men
to bring her back from Fogo, and I agreed to
give 6*l*. for the whole trip to that island. We
gave a passage as far as Cape Freels to a young
woman whose father lives in the Gooseberry
Islands, and supports himself and his family entirely by farming, raising vegetables and beef
and mutton for Greenspond and the neighbourhood. This is one of the very few instances in
the island of a settler entirely independent of
the fishery. We sailed by some low islands and

many dangerous sunken rocks down to Cape
Freels, where we dined; and I then sent the
boat on to Cat Harbour, as one could easily get
alongshore for the seven miles to that place.
The land for some distance on each side the
cape is very low, and composed entirely of granite, with here and there a little gneiss, and
as the water deepens very gradually out to sea
there is a succession of sand beaches for many
miles: the low land, shallow water, and sand-
beaches being unexampled on this side of the
island, except in this locality. The country
was generally barren, but the comparative ease
and freedom with which one could walk over it
made it most delightful, and in these seven
miles I got more pleasant exercise, and felt,
therefore, greater energy and elasticity of spirit
than I had done for weeks. There were in
some places ranges of sand hills twenty or
thirty feet high, with low marshes behind
them; but as we approached Cat Harbour, we
rose on to a fine granite bluff, not so striking
from its height, which was inconsiderable, as
from the great blocks and wide unbroken sheets
of granite that stretched about it. Arrived at
Cat Harbour, we found the boat ready and the

men waiting for us at the house of Mr. Gibbons, one of the principal settlers in the place.
Cat Harbour does not afford much shelter, the
Harbour being formed by an island, and the
entrance on each side narrow and dangerous.
It is, however, an excellent place for the fishery,
and in the summer is crowded, though in winter nearly deserted. As the midship-room was
now fully occupied, I slept astern, with my
head in the cuddy and my legs out, expecting
to sail before daylight.

August 25th.—On waking found a heavy
gale blowing from the north-east, with clouds
and fog, the harbour all in a foam, and our
departure impossible. A boat was sent off with
some difficulty from Mr. Gibbons', to fetch
us ashore, and I breakfasted and spent the day
there. In the afternoon the weather cleared up,
and I went some distance over the marshes into
the country, but saw nothing worthy of notice.

August 26th. — At dawn we were under
weigh. We sailed alongshore as far as Deadman's Point, where the sand beaches ended
and a rocky shore began, and then, passing by
some low rocks called the Penguin Islands,

sailed through the islets called the Wadhams.
There was a large island of ice aground off
these islands.   Penguins were formerly so
abundant on these shores, that their fat bodies
have been used for fuel: they are, however,
now all destroyed, and none have been seen
for many years.   There are three low rocks
thirty miles out to sea from Cape Freels, called
the Funk Islands, whence several boat-loads of
sea-birds' eggs are annually brought away.   It
was a lovely day to-day, but when we were off
Cape Fogo it fell calm.   A light breeze, however, sprang up in the evening, which carried
us round, and we entered Fogo Harbour about
seven o'clock.   The northern side of Fogo
Island is very bare and rocky, with lofty headlands rising perpendicularly from the sea.
The harbour is excellent when it is once entered, but a string of low islands stretching
across its mouth compels vessels to come in
by narrow channels on each side of them.
The eastern entrance is very narrow and rather winding, while the western entrance is at
the foot of a bold precipice, called Fogo Head,
500 feet high, where the winds necessarily

baffle and vary, unless blowing right in. The harbour was surrounded with houses, two of which were large and prettily situated. According to the prevailing custom in an out-harbour where there are no inns, I might have selected either of these houses for my temporary residence; but, as I had been informed at Greenspond that one gentleman had visitors, I went to the other house, where an agent of Mr. Slade's resided, and he received me with all imaginable kindness and hospitality.

August 27th and 28th.—A gale of wind from the north-east threw up a tremendous sea on the islands at the mouth of the harbour. A few miles north of Fogo is a cluster of small islands and rocks of a very dangerous character, and the sea all round contains sunken rocks, shoals, and small islands, which render its navigation at all times perilous to strangers, and in rough weather dangerous even to those best acquainted with it.

August 29th.—Hired a small boat and a couple of men to take us on to Toulinguet. We had some difficulty in beating round Fogo Head, the sea still running high, but we succeeded, and ran up to Hare Bay, where I wanted to examine some red rocks, which I found to consist of sienite. We then ran down to Change Island Tickle, a long narrow passage between two islands, where vessels often anchor when they cannot get into Fogo Harbour. Here there were a good many houses scattered about, but we beat through without stopping. In the space between this and the New World Island a heavy sea was rolling in, dashing the surf over several sunken rocks, and two or three large lumps of ice were whirling about in it. When off Herring Neck it fell calm, and we took to our oars, and slowly rowed in among some lofty cliffs, until it became so dark we could scarcely see: we then landed in a small place called Ship Cove, where a family resided, by whom we were hospitably received and entertained.

August 30th.—On coming out of doors this morning we observed that the house was strangely situated at the head of a little cove or crevice in the rocks, with lofty perpendicular cliffs on each side of it. Many other houses were scattered about the shores farther on, perched in the recesses and numberless nooks

and corners of this singular place. Narrow
lofty necks, or rather walls of land, principally
composed of dark slate rock, were indented
and separated from each other by equally
narrow and deep inlets, and without the chart
I should have been completely puzzled to
know where we were or how we got there.
The air was thick with smoke proceeding from
fires which were said to be raging in the woods
near "Seldom-come-by Harbour," and on the
south side of New World Island. We had
now a fair breeze, with which we sailed round
French Head, and into Toulinguet Harbour.
Here I threw myself at once upon the hospitality of Mr. Slade, who received me very
kindly and entertained me most hospitably for
the next month. To Mr. Slade and Mr. Peyton,
indeed, I am so much indebted that I cannot
help thus publicly acknowledging it, although
I was also greatly obliged to several other residents of Toulinguet, who kindly helped to
render my stay there very agreeable.

August 31st to September 3rd.—Mr. Slade
got ready a little cutter which he had fitted up
as a small yacht, having two berths and a comfortable cabin amidships, and he and Mr. Pey-

ton determined to accompany me on an excursion to the river Exploits.   On the 2nd of
September the smell and smoke from the
fires to the south of Toulinguet increased so
much, and the bright light at night was so
vivid, while small ashes were falling about,
that it was feared the fire might reach out to
the harbour and consume it.   We heard also
of fires burning in Loo Bay, and on the shore
near Cape St. John.   Some of these fires
might be caused by lightning, but they must
chiefly have arisen from carelessness, fishermen going ashore and lighting a fire, either to
boil their tea or keep off the mosquitoes, and
not extinguishing the embers.   The long series
of warm weather had made the woods quite
dry, and the mosses and lichens hanging on
the trees are like so much tinder.   The bark
of the birch-tree flares immediately, like paper
dipped in turpentine, and is always used for
lighting a fire when it can be got; while the
barks of the fir tribe are highly resinous.   A
great quantity of the bark or "rinds" of trees
is always used to spread over the fish-piles,
and men are employed regularly during the
summer, in some places, to go into the woods

and strip the standing trees. This they do by
beating the bark with a small bat, and then
making a perpendicular incision and two horizontal circular cuts at each end, when the bark
peels off. These men of course light fires,
which, unless extreme care be taken, spread
along the ground through the crisp moss,
and ultimately fire the woods. This practice of
" rinding" is in other respects very pernicious,
as it checks the growth of and destroys hundreds and thousands of trees, that might ultimately become pretty fair timber if left alone.
From this cause, and from all the tolerable
timber being continually and wastefully cut
down, the woods near the sea-coast and in the
accessible parts of the country are not equal
to those in the interior.

September 3rd.—Having sent the cutter
round to Back Harbour, we sailed this morning, although the wind was against us, and
beat up to the neighbourhood of Morton's Harbour, when it fell calm, and Mr. Peyton left
us, going on in his skiff to Exploits Burnt
Island, where he had some business to trans-
act. At dusk a light air sprang up, which
took us nearly to Exploits Burnt Island and

Slade and I, with Tom and Simon, took a four-oared gig we had brought with us, to go in search of Mr. Peyton. The men told us to seek for a narrow passage between two islands, on entering which we should see the houses. It was now quite dark, and on getting under the cliffs we were involved among rocks and passages, and could find no inlet. The water was more beautifully luminous than I ever saw it before. Every dash of the oars and surge of the boat sent waves and bands and eddies of trembling phosphoric light circling and glimmering around us, and dimly lighting the dark rocks and cliffs over our heads. We returned on board, and found the men had sent us to the wrong place, the harbour being half a mile farther up. We accordingly stood on, and then sent them in the boat, while we managed the cutter, standing off and on about the harbour. About one in the morning they came back, with no news of Mr. Peyton; so concluding he had gone on with his skiff, we also sailed up the bay.

September 4th.—Heavy rain and stormy squalls of wind during the night. When I came on deck, I found we were beating up against wind and tide in smooth water, through various channels among the islands of which this bay is full. The land both of the islands and the main was moderately high, with steep cliffs and deep water close in shore. The tops of the cliffs were covered with wood. About 4 P.M., after many fruitless endeavours to get through a chain of islands on the south-east side of Thwart Island, we were obliged to anchor in a small cove close by, till the strength of the tide current moderated or the wind shifted. The channel on this side of Thwart Island is not very deep; but on the other side, although in some places scarcely a mile in width, no bottom has been reached in the middle of the channel with 120 fathoms, and 90 and 100 fathoms are the depths close in-shore. What narrow and precipitous ravines must these be, and how were they produced? What a singular sight would it be on dry land to see a long winding valley scarce a mile wide with perpendicular walls rising to the height of nearly 800 feet! The discovery of such a ravine might induce the geologist to speculate on the amount of erosion necessary to produce it, but in this case erosion can hardly be supposed to have been the agent; for the phenomena seem rather to be the result of some modifying cause, acting during the period when these slate rocks were first elevated. Great irregular faults must, I think, be necessary to produce such a structure, and this supposition is borne out by the fact, that most of the deep inlets of Newfoundland have the same general bearing as the prevailing strike of the rocks, namely, about north-north-east and south-south-west (see Report). It is singular, however, to see in these narrow land-locked inlets a greater depth of water than is to be found in any of the seas within two or three hundred miles of the coast of Britain.

September 5th.—Continued our progress, beating up against a south-west wind, but with the current more in our favour. Landed for a short time at Lower Sandy Point, where there is a salmon fishery of Mr. Peyton's, but the men there had seen nothing of him; and about 4 P.M. we came in sight of Upper Sandy Point, and the mouth of the river Exploits. In tacking for the mouth of the river, we stood a little too far in on the south side, and ran the cutter aground. There are some hills, called the

Shute Brook hills, visible up the valley of the
river, which form the mark for the entrance of
the channel; and had we kept these always
open, we should not have gone beyond the
deep water. The sides of the inlet here are
very shallow, the bottom being mud and coarse
sand. A boat put off to us from Upper Sandy
Point, in which we were happy to find Mr.
Peyton. He was in Burnt Island harbour
while the men were looking for him, and in the
only house they did not visit. He had then
beat up all day and night in his open skiff
through the rain, had passed us, he supposed, in
a squall of rain while we were at anchor, and
sailed by Lower Sandy Point in the dark. We
landed, and found a very pleasant comfortable
house, where Mr. Peyton used formerly to reside, having an excellent garden behind, with
a grass-plot and a few scattered birch-trees
between it and the river in front, and altogether a very pretty looking and quite an English sort of place. The mouth of the river here
is 400 yards wide, with water sufficient for a
schooner, when in the proper channel. The
tide runs up about five miles farther. The land
on each side is low and flat, with a steep ab-

rupt ascent, about a mile from the present river, on each side. These ascents were probably the old banks of the inlet, when it stood at a lower level than now. The whole was covered with wood, stunted and thick.

September 6th and 7th.—After a long continuance of the most beautiful weather, it broke at last just at the very time when I most wished it to be fine. A perfect deluge of rain now poured down, making it madness to think of starting. Mr. Peyton entertained us with discoursing of the Red Indians. He had frequently seen them, having found them on the Red Indian Lake and elsewhere. He had captured one of the women, who was taken to St. John's, and who lived some time with Mrs. P. as a servant. He described them as a fierce and savage race, supporting themselves entirely by hunting and fishing, and forming their wigwams not of bark, like the Mic-Macs, but of skins. These wigwams were raised on wooden platforms, which, together with some other structures intended apparently for storehouses, were formed with much skill. They seem to have had many peculiar manners and customs, the record of which is now probably lost for ever. Among

their most prominent customs was that of
smearing their persons and implements with red
ochre, as also their dresses, which were formed
of deer-skin.   Many years ago they were very
troublesome to the European settlers, frequently
stealing boats, nets, and implements.   One
night Mr. Peyton, being about to set out for St.
John's, had a boat ready loaded down at Sandy
Point, containing sundry articles that made a
rather valuable cargo.   His bed being packed
up on board, he did not lie down, but walked
in and out of the house during the night, keeping watch.   Once his attention was caught by
a dark object lying on the beach at some little
distance, and waking one of the men, he asked
h im whatit could be ; the man replied, it was
probably a splitting-table* he had left there in
the afternoon.   Between three and four o'clock
Mr. Peyton began to feel tired, and laid down
for an hour in the house ; and on going out
again, he found his boat was gone, and that the
dark object on the beach had likewise disappeared.   He at first thought the boat might
have got adrift ; but on examining the ropes by

* A table on which the salmon are split, previously to
their being salted.

which she had been moored, he found them
cut with some sharp instrument. He immediately sent one punt up the river, and went
himself down the bay, where he found his boat,
the next afternoon, stove in on the rocks and
plundered. Going to the valley of an adjoining brook, he found there a wigwam covered
with the sails of his boat ; the cases of a gold
and silver watch ; a broken pistol, and various
other articles. A woman, whom he afterwards captured, confessed that she and another
woman, with two men, were the parties concerned. A man, perched in a forked birch-
tree close by, had watched him all night long,
and, taking advantage of his prolonged absence,
when he laid down, had cut the boat loose and
turned it adrift. The dark object on the beach
was their canoe, with which, when his boat was
adrift they paddled off to her, and towed her
away. Mr. Peyton assured me that neither he
nor his men had ever injured the Red Indians,
though, had they chosen, they might easily have
destroyed them all, and that the Red Indians
in turn, apparently conscious of being in his
power, never attempted any personal injury
towards them, but still considered all plunder

that came in their way as fair game. Before his time, however, when the Red Indians were more plentiful, and the settlers fewer, the former were often shot, and though, perhaps, sometimes necessarily in self-defence, they were, too often, no doubt, wantonly persecuted, and their depredations on property visited with wholesale and indiscriminate revenge and destruction. Their destruction, however, was not wholly due to the English, the French had a still greater hatred of them, and contempt of their lives, which they even to this day preserve. Their very term of " sauvages " for all whom we call Indians denotes this. The Mic-Mac Indians were, however, the most efficient instruments of their destruction ; and according to the account which an old Mic-Mac Indian gave to Mr. Peyton, the first enmity between the two races arose in this way. When the Mic-Macs first visited the country, they and the Red Indians were friendly. About a hundred years ago, however, the French offered a reward for the head of every Red Indian. To gain this reward, the Mic-Macs privately shot some of them; and one day, in descending a river near St. George's Bay, they fell in with a party of them, while they had the heads of some of their nation concealed in the canoe. The Red Indians invited the Mic-Macs ashore to a feast, during which, some children playing about discovered the heads! No notice was taken till each Mic-Mac was seated between two Red Indians, when, at a given signal, these two fell upon him, and slew him.* After this, they fought at the north end of the Grand Pond, and at Shannoc brook on the Exploits river, and, indeed, wherever they met. In these encounters, from their possessing fire-arms, the Mic-Macs were victorious. Mr. Peyton said, the Red Indians had a great dread of the Mic-Macs, whom they called Shannoc, and used to point to Shannoc brook, on the Exploits river, as the way by which they arrived in their country. The woman, who lived with him some time, was greatly alarmed at the sight of two Mic-Macs who came once to visit him, and hid herself during their stay. They were acquainted with another tribe of Indians, whom they called

---

* A confused account of this same feast was given me in St. George's Bay, by Sulleon, describing the Mic-Macs being placed between two Red Indians, but saying nothing about the story of the heads.

Shaunamunc, and with whom they were very
friendly. These came from the Labradore,
but were not Esquimaux, whom the Red Indians also knew and despised for their filthi-
ness. The Shaunamuncs were dressed in
deer-skin, and not seal-skins, but their deer-
skins were not reddened. They answer, I believe, to the Indians called Mountaineers, on
the Labradore shore. The Red Indians traded
with these Shaunamuncs; receiving stone-
hatchets and other implements from them, and
they mutually visited each other's countries.
This fact in some measure corroborates the supposition, that the total disappearance of the Red
Indians, for the last ten or fifteen years, is not
due to their utter destruction, but to their having passed over to the Labradore. Mr. Peyton
said he had heard of a body of strange men in
red deer-skins having been seen on the Labradore coast; and the same occurrence is mentioned in Sir R. Bonnycastle's entertaining
book on the Canadas. That there are any
Red Indians left in Newfoundland, now the
coast is become so much settled and the Mic-
Macs so frequently traverse the interior, is in
the highest degree improbable for a hunting

tribe is necessarily migratory and widely dis-
persed and cannot be secluded in any corner of a country, without either themselves,
their implements, or traces of their encamp-
ments being necessarily seen.   Mr. Peyton re-
members the deer-fences of these people along
the river Exploits, and gave me the following description of them.   There was a series
of stockades of trees for thirty miles along
the river, the trees having been cut down to
about breast-high, and the fallen parts piled
and interwoven among the stumps.   At intervals, in these stockades, lateral avenues went
off some hundred yards into the woods, gradually widening as they proceeded from the
narrow passage through the stockade.   The
ends of these avenues or passages, on opposite
sides of the river, were never opposite to each
other, but placed alternately, so that on the deer
coming out to the river, they must either go
up or down a considerable distance before they
found another avenue by which to proceed.
Places were made in these stockades for men to
watch, and pieces of birch-bark were tied loosely
here and there to flutter in the wind and attract
the attention of the deer.   The men then either

shot them with arrows, or rushed in a body into the river, and speared them, giving chase in their canoes in the deep parts.

September 8th.—It gradually cleared off this morning, and about 11 A.M. we set off in a four-oared gig and two small "flats," these latter being little flat-bottomed boats with square ends, about the shape of a common knife-tray. We rowed the gig up for about five miles, the river gradually becoming narrower and more shoal. Here we found a small ledge of rock running across the river, forming a little face at low-water, but covered at high-water, which just reaches its level. Above this no tide extends, and the river is rapid and shallow. We left the gig at this spot, and set out, our party being Mr. Slade, Mr. Peyton and his son, and myself, together with two of my men, and three of theirs : we were thus two to each flat, and five to hunt for game, and tow the flats at the rapids. We stopped among some rocks a little way up, and shot a couple of seals, numbers of which were playing about. They sank, however, immediately, and, owing to the great rapidity of the current, we could not get up the bodies. About a mile above the first rapid is the

Bishop's Fall, so named from the present bishop
of Nova Scotia having visited the place.   This
is a very violent rapid, of 150 yards in length,
where the river rages in narrow and tortuous
channels worn in a ledge of hard slate-rock
twenty feet high, which here crosses its course.
We were obliged to unload the flats, and carry
first the provisions, and then the flats themselves, over ledges of sharp rock, and up a steep
woody bank, and launch them again above the
rapids.   Though small, from their necessarily
stout build these flats were rather heavy, each
requiring our united strength, when slung upon
cross-poles, to lift and carry.   To drag them
would have been impossible, as, from the sharp-
ness of the slate-rocks, they would have been
cut to pieces.   Above the Bishop's Fall there
was a "steady," for a little distance, and then
a succession of smaller rapids, frequently obliging all hands to wade out, and lift, haul, and
push the flats over the rocks and ledges.   The
water-course was generally 200 yards wide, but
not more than half that was usually occupied
by the water at this season of the year ; the rest
of the bed of the river being composed of bare
rocks and boulders, and here and there banks

of pebbles.   The banks of the water-course were
generally sloping, rising fifteen feet above the
present water, and composed sometimes of
small slate-cliffs, and sometimes of sand or clay
resting on the slate, the whole being covered
with coarse gravel and great boulders.   Immediately on the top of the banks grew the
dense wood, the lower parts of the trees being
scarred and bared, and sometimes a number
of them torn up and laid prostrate.   This,
Mr. Peyton told me, was the effect of the ice
in spring.   The water, when the snow begins
to melt, is penned in and dammed up by
great blocks of ice and frozen snow at the
rapids, and at the ledges of rock which cross
the river : it accordingly invades the woods, and
when the icy barrier gives way, it rushes along
with great fury, drifting great blocks of ice, and
tearing everything along with it.   Our progress
up the rapids was necessarily slow ; and those
who walked had often to wait for the boats.
Just as it got dark, we reached a spot about
half way to the great falls, where there was a
tilt in the woods.   The boats came in about
half an hour afterwards, by the light of the
moon ; and we bivouacked for the night, sup-

ping on some shell-birds we had shot as we
came along.

September 9th.—On setting off, Slade and
I went across to the south side of the river,
which is hereabouts pretty free from rapids. We
soon reached the mouth of the Great Rattling
Brook, a considerable stream coming down
from the north. Its character may be known
by its name, "rattle" being the term used
in Newfoundland for "rapid." After wading
breast-high through the holes, and flounder-
ing in the currents, at the mouth of this brook,
we found, that had we gone up it a little way,
we might have leaped across dryshod, jumping
from boulder to boulder. Half a mile above
this, the walking became so very bad, that we
crossed to the north side by the aid of one of
the flats, where we did not find it much better.
Here began a succession of very bad rapids extending for two or three miles, and costing us
more than as many hours to master them. The
walking even in the dry part was horrible,
being a succession of stepping and jumping
from one round slippery stone to another, or
over still more slippery and sharp edges of slate-
rock. It was intolerably fatiguing to the feet,

making them constantly ache with pain. The
men, in addition to this slippery work, were all
the while above their knees in water; and on
going to help them, I found I did more harm
than good, as it was quite as much as I could
do to keep my footing; and even that was
out of the question when pulling at the line.
Toiling steadily in this way, we succeeded,
about two o'clock, in reaching Buchan's Island,
at the bottom of the ravine of the great falls.
Buchan's Island is a lofty rock standing in the
middle of the river, at a point where it makes
a sudden turn towards the north. The corner
where we stood, being the inside of the bend of
the river, was heaped and piled with an enormous assemblage of great angular masses of rock.
The tall rock of Buchan's Island, and the bare
precipitous ledges on each side of the river, mark
the original site of the falls, which, in the lapse
of ages, have cut back a mile and a half to their
present position. The ravine, through which
the river flows, between Buchan's Island and
the falls, is about fifty or sixty feet deep, with
banks, for the most part, rocky and precipitous, but broken occasionally, and here and
there presenting the appearance of a worn

and ruined wall, extending across it, eaten
into fantastic shapes, which mark the site of
some band of stone harder than the rest, and
which the falls have not so completely succeeded in destroying and sweeping away. The
bases of these ruined walls the river is still
assailing, fretting over them in fierce rapids,
with deep sullen whirlpools below.   On a bare
sheet of rock near Buchan's Island were two perpendicular circular holes, like wells, two feet
across and six or eight feet deep.   They were
exactly circular, nearly full of water, and each
contained a large stone and a quantity of sand at
the bottom.   There were several similar holes
of shallower dimensions ; being all caused by the
whirling round of the stone from the rapid currents of water rushing over them.   There was
formerly a good path through the woods, used
by the Red Indians in order to avoid the falls ;
and it was our object to find this, and carry
over our provisions and one boat, that we might
proceed up the river.   The path, however, was
now overgrown, and we could not find it ; but
we struck into the woods, each person with
a heavy load on his back, thinking that as we
had the roar of the river on our left to guide us,

we should hardly be likely to miss our way.
We walked and scrambled along through the
woods, directed by Mr. Peyton and his men,
who were experienced hunters, staggering
under our heavy burdens, over the mossy rocks
and among the thick bushes, but still keeping,
as we thought, a pretty straight course.   About
six o'clock, we still heard the river on our left,
and thinking we must have come at least a mile
and a half, we struck out for it.   On getting
out of the woods, we stood on the brink of a
ravine, at a spot which, after a little attention,
we recognised as being about 100 yards above
the place we started from.   We were all knocked
up, and at the disheartening discovery of our
having only gained 150 yards in two hours' intense labour, we gave it up for the night, threw
down our burdens, and proceeded to encamp.
One of the men, named Blake, had strayed and
was missing; Mr. Peyton, and one or two
others, went in search of him and of the path,
while I selected a tolerably level spot near a
little brook for our bivouac, and proceeded to
light a fire and pile boughs for our shelter.
Soon after dark, Mr. Peyton came in, having
found neither the man nor the path: he had,

however, heard the man's voice at a distance,
and as he was an old furrier, though he had
neither gun, flint, steel, nor hatchet, he said
he would manage to pass the night and join
us in the morning.

September 10th.—Blake came back while
we were at breakfast, having found his way
out to the falls last night, hit upon the path this
morning, and traced it back to about 100 yards
beyond our bivouac. We accordingly shouldered
our packs, and set off to the end of the path.
The entrance was overgrown with bushes, but
as we proceeded we could distinguish traces of
it. Setting off in a hurry, we wandered again,
and found ourselves at fault, searching in vain
for our path, and Mr. Peyton went off by himself in one direction, while we sent Blake back
to the river to make another attempt. This
he did, and found the path within fifty yards
of the place where we were sitting. We struck
out accordingly, but had scarcely gone another
hundred yards when we again lost ourselves, but
once more found the path proceeding through
some alder bushes which had grown over it.
Beyond these, the ground got rocky, and we
could more easily discern our way, which was

then in some places five feet broad. When the
ground is covered with moss, it is almost impossible to trace an old wood-path ; and to inexperienced eyes there is no sign of any path whatever. Practice, however, I found had given me
almost as much acuteness of vision as the old
hunters, and I could follow a " footing " of a
man, or detect that of a deer, where formerly I
should not have been able to see any mark at all.
Passing through much thick wood, and over
a small marsh, we came out at a spot 200
yards above the falls, where the river had just
the same shallow and rapid character that it
had below Buchan's Island. Peyton came soon
after, having been led off in chase of a fine buck
he had seen in the woods, and of which we observed the fresh traces on the little marsh we
had just crossed. We now divided : Slade and
I went to examine and sketch the falls; Mr. Peyton, his son, and one man, went up the river to
examine into its state ; and the rest went back to
clear the path, and bring forward more of the
provisions. The river above the falls is about
200 yards wide ; shallow, rapid, rocky, and
full of boulders. It falls by a succession of
small leaps, in various channels, twenty or
thirty feet, and then by one great leap, in two

channels, with a small woody island between them, thirty or forty feet more. The surrounding rocks are composed of hard, fine-grained gritstone, of a brick-red colour, and of slate-rock ; the beds striking obliquely across the river, and dipping at an angle of 45° to the north-west, or against the stream.   In consequence of this position of the beds, the water foams and frets over their edges, and falling on the inward-sloping ledges, spouts up into jets of spray, boiling and thundering over the many obstructions to its course.   Beneath the present falls is a whirlpool about fifty yards across, below which the water is confined to a channel not twenty yards wide, through which it shoots with great velocity, and again expands into a pool ; and so on, throughout the ravine, down to Buchan's Island.   On each side of the water, at the time of our visit there, was a space of fifty yards of bare rugged rock, and small precipices, whose angles were inconceivably numerous and sharp, being the space over which the river extends when it is flooded.   Immediately on the brink of the river-course, the thick woods commence, and stretch away to an indefinite distance into the country.   About a quarter of a mile above the falls, the channel

of the river suddenly turns to the left, re-
suming its westerly direction, which it pre-
serves for the remainder of its course.*

In the afternoon, Peyton not having returned, and the clouds threatening rain, we
set to work to make a bough tilt.   Cutting
down a stout pole, we stretched it between two
trees at a height of about eight feet from the
ground.   Then getting a number of smaller
poles, we rested them side by side in a sloping
position against the windward side of the cross-
pole.   Weaving a few boughs through these
slanting poles, we cut a great number of
branches of fir and spruce, covered with leaves,
and, beginning at the bottom, laid them one
over the other with the leaves outwards, in
tile-like fashion.   This formed a roof imper-
vious to rain, and blocking up the sides with
heaps of boughs and moss we formed a kind
of weather-tight hut or shed, open only in
front.   On this side we made a great fire.
The whole construction did not cost us more
than an hour.   Peyton and his party did not
return till after ten at night, and were then

---

* For further details on the rocks and falls of this spot,
see the Report on the Physical Geography of the Island.

too tired to do more than drink a basin of tea and lie down to sleep.

September 11th.—We now called a consultation. Mr. Peyton reported that his party had, with great difficulty, although without a load, got about eight miles farther up, having sometimes to go through the wood, sometimes to crawl over rocks, and sometimes to wade through mud. He had not before been up the river above the falls except in winter, when, all being frozen, it is comparatively easy. He declared that while the water was so low it would be nearly impracticable to convey the flat for these eight miles, though above that the water seemed deeper; and in addition to these difficulties, no deer had been seen, so that we should have to carry provisions. To bring up the flat to where we were, a mile and a half through the woods, we knew would take half a day, and to carry it the eight miles would require a whole one; to get on a dozen miles farther, therefore, would require two days. Even then we should be twenty or thirty miles from the Red Indian Pond, so that to get the flat into the pond would apparently require five days: before which time our provisions would be exhausted.

On the other hand, to reach the pond without the boat would be almost useless, and to carry six or eight days' provisions on our backs impossible. Besides, the shoes belonging to most of the party were nearly worn out, being destroyed by the water and the sharp rocks, and only one or two of us had a second pair. Had Mr. Peyton's party succeeded in killing a deer we could have depended on that, and advanced two or three days journey at least; but to carry our provisions only a few miles farther, and then come back again, was useless. Moreover, Mr. Peyton and Mr. Slade were obliged to be at home about the 15th, to meet the judge and attend the court. As far as regarded the geological part of the business, I was perfectly satisfied that there was no cone near us, and from Mr. Peyton's account of the rocks, there appeared no chance of any in the neighbourhood of the pond. The flat country which I had observed north of the Grand Pond did not therefore extend in this direction, but ran out more towards White Bay. For all these reasons we determined to return, and I meant to proceed by sea farther along the coast. To any future explorer who is desirous of going to the Red

Indian Lake I should recommend rather to trust to a couple of Indians and a bark canoe, than to any number of furriers and their heavy flats. And from Mr. Peyton's account I am inclined to think that an easier route might be found than ascending the river Exploits from its mouth, by going from Badger Bay to some large ponds in its neighbourhood, and descending a brook that flows into the river Exploits at a point about ten miles above the falls.

Having determined to return, we hung up in the tilt a "nunny bag"* full of bread, and hid a quantity of shot to serve for the furriers on a future occasion, and carried the rest back through the wood to the flats. Peyton, Slade, and Blake took one flat, and young Peyton, Simon, and I, the other, while the three men walked. Just as we pushed off it began to rain, and continued in heavy showers all the day. In our flat Simon sat amidships with a pair of sculls or short oars, while John Peyton stood in the bow and I in the stern, each with stout poles to guide and " hold up"

---

* This is a bag made of seal-skin converted into a knapsack: what the origin of the word " nunny " is I cannot tell, but it is in universal use in the country.

the flat in the rapids. Sculling, and poleing in
the deeper parts, we every now and then shot
down a rapid, picking with instant decision the
best channels among the rocks, which are always
those where there is most foam and the waves
rage with greatest violence. One or two of the
rapids were a mile long, and in passing these it
requires the polemen to look out carefully, for
as you dart along you have every moment to de-
cide on which side of the rock it is best to pass,
and which channel it is best to slip into, avoiding promptly the dangers close at hand, and
looking out at the same time for those which
are to come. It is requisite to preserve steadi-
ness and coolness in the midst of the roaring
and flashing of the waters, and never to hesitate
for an instant. The moment the bow of the
boat catches upon a rock, the stern poleman
must fix his pole in the right position, and hold
up the stern of the boat against the stream till
the bow is freed, for, if she once swings broadside
to the stream and catches between two rocks,
the water would pour over her side, or roll her
over and capsize her in an instant. Should
such an accident happen, not only is all the
cargo lost, but even the lives of the crew are

endangered, since, if a man once falls in these rapids, even where they are not more than knee deep, it is very difficult for him to regain his footing. We were young hands in our flat, but we several times passed the old ones, and only stuck fast twice, and Blake the old furrier awarded us the praise of managing it as well as if we had been used to the brook for twenty years, on which we plumed ourselves accordingly.

On reaching the Bishop's Falls, we had carried over the baggage and one of the flats, when the walkers overtook us, and helped us over with the other. I then walked down to the lowest rapid, where we had left the gig. As we went along I shot two seals, but was not able to get to them; and we all embarked on the still waters of the deep part of the river just as it became dark. When about half way down, the full moon rose among black drifting clouds and squalls of rain, and I was gratified by seeing for the first time in my life a distinct lunar rainbow. It had a pale light, but was clearly visible against a dark cloud.

September 12th.—Very warm, with heavy showers. Mr. Peyton was residing at his

house at Upper Sandy Point, when Mr. Cormack made his second excursion into the country with two Mic-Mac Indians. In this excursion he went in at Hall's Bay, visited the
banks of the Red Indian Lake, and came out
down the river Exploits. Mr. Peyton said he
saw Mr. Cormack when about to depart from
Toulinguet; and when, about a fortnight
afterwards, he came to his house from the
country, he was at a loss to recognise him.
They had been unsuccessful in hunting, and
for the last three days had had no food. Pale
and emaciated, Mr. P. said he could scarcely
have believed a fortnight could have wrought
such a change in any man.* Some years ago

* I have mentioned, and insisted very strongly, in
this journal, on many circumstances unimportant in themselves, but which, I thought, would aid in conveying
some notion of the difficulty of penetrating the interior
of Newfoundland. I have done this, because when I first
came to the country I laughed at the idea of this extreme difficulty, and could by no means conceive its existence. I had, indeed, never seen anything which could
give me an idea of the character of the country, and I
therefore rashly concluded the difficulty to be more ima-
ginary than real. After a few trials, my ideas, however,
were enlarged, and perhaps took rather an opposite bias,
magnifying the impediments rather than diminishing them.
This was natural, and in the course of this summer I some-

a man-of-war had been stationed at the Bay of Exploits, and in the winter a party went in to explore, traversed the Red Indian Lake and the neighbouring country. One of the Lieutenants, assisted by Mr. Peyton, made a map of all the country explored; but what became of it he could not tell.

September 13th, 14th, 15th.—We were detained in various parts of the bay by fogs and north-easterly winds with heavy rain, and it was not till six o'clock in the morning of the 16th that we entered the harbour of Toulinguet. As we sailed in, two brigs made their appearance, and came in together shortly afterwards. These belonged to the house of Slade and Co., and the circumstances of their late trip were curious. They sailed from Toulinguet exactly seven weeks ago, for Lisbon; never saw each other during the voyage, but one entered that port only two hours before the other.

times got on better than I previously believed possible. In fact, however, after a year's practice the difficulties were diminished to me : I had become accustomed to treading on the marshes and threading through the trees, and my gait and movements were become more Indian-like, and accommodated to circumstances. Without about a year's practice, the best walker and the most active man in the world would lose all his superiority in the wilds of Newfoundland.

They unloaded their cargoes of fish, sailed again together, and without keeping together intentionally, but each making the best of her way, the one which was two hours later into Lisbon was just two hours before the other off the harbour of Toulinguet, when her Captain hove to for the other to join him. We found the judge's vessel at anchor in the harbour, with the judge, solicitor-general, sheriff, clerk, and constables on board. They had arrrived a day or two after we left, and the judge said he had had a great mind to run on and join us in our excursion up the river. There was a brilliant aurora visible to-night, of precisely the same appearance that I have described before, except that the waving circular band of light struck me as appearing double, as if there were two parallel bands of perpendicular rays, the one seen partly through the other. The progressive motion of the light was also in this case from the north-west to the north-east, instead of being as usual from the north-east to north-west.

September 17th to 30th.—I remained during this time at Mr. Slade's, partaking of the hospitality of the various residents on the occa-

sion of the judge's visit, and I was amused
by attending the court. The delay, although
pleasant enough, was involuntary on my part.
Mr. Slade and Mr. Peyton expressed an inten-
tion of accompanying me farther west as soon
as the business of the court was over. I
could not run away with Mr. Slade's cutter,
at a time when he might himself require
it, and there was no other vessel to be had.
Indeed, even had there been a craft ready, I
should scarcely have been able to proceed. The
wind blew steadily from the west, shifting now
and then only to south-west or north-west; and
a constant current sets alongshore from the
north and west. For these reasons beating to
windward would have been a very slow and
tedious process for small craft; and I was
therefore compelled to remain at Toulinguet.
The court-house of Toulinguet consisted of
one good-sized room, with apartments for the
gaoler, and a cell or two below. A chair
elevated on a platform of boards, with a table
before it, was the seat of the judge. A table
on the floor was set apart for the clerk of
the court, and there were a few chairs placed
round it for the use of the sheriff and the

barristers; a bench along one side of the
room was reserved for the grand jury, and a
similar one opposite for the common jury. If
the latter wished to consult as to their verdict,
they were led out of doors by the constable,
and assembled on a rock close by, where
they were locked up—in imagination—till they
agreed. There were no cases of any consequence, the most serious being one in which
a boy had (accidentally, as it turned out) shot
another. The other cases were either actions
for debt, at the suit of the merchants, or between the planters themselves, or actions arising
out of trifling trespasses and disputes. Some
of the addresses to the court, when the plain-
tiffs or defendants acted as their own counsel,
as well as one or two of the verdicts of the jury,
were sufficiently ludicrous, and caused afterwards great merriment, but they would lose
all their humour unless accompanied by the
voice and action, and the simple earnestness of
the speakers. The judge's vessel was a mer-
chant-brig hired for the purpose, and fitted up
fore and aft with cabins and apartments for
the various law-officers, from the judge down
to the constable. The coast is divided into

two circuits, the northern and the southern.
The courts on the southern circuit are held
at Ferryland, St. Mary's, Placentia, Buria, and
sometimes in Fortune Bay or Cape la Hune.
The northern circuit comprehends Harbour
Grace, Trinity, Bonavista, Toulinguet, and
sometimes Greenspond or Fogo. What would
an English judge think of being shipped off,
with all the law-officers, barristers, lawyers,
clerks, and constables, and sent cruising over
rough seas and along wild shores for a month
or two every year?

September 30th.—We sent the cutter round
to Back Harbour to be ready for a start. As
soon as she got outside the main harbour,
however, it fell calm, and she was drifted to
leeward : light airs assisted her a little, but she
was at one time ten miles to leeward, and did
not get round till late the following evening,
being thirty hours in getting from one harbour to the other, or about three miles to
windward.

October 5th.—Light westerly airs still continued, and the example I had just had showed
me how impossible it was to get to the westward
with them. As, however, the schooners going

to St. John's were all about sailing, and I was afraid, if I delayed too long, that I might lose the chance of a passage, and be detained here all the winter, I determined to make an attempt, and Mr. Slade kindly lent me his cutter, neither himself nor Mr. Peyton being able to go. At 3 p.m., therefore, favoured by a light wind from the south, we pushed out of harbour, and made for the Bay of Exploits. The wind soon shifted again into the south-west, and, aided by the tide and current out of the bay, drifted us rapidly off to sea. When we came to tack, we found the cutter had somehow got out of her proper trim, and we could not get her to "stay;" accordingly, as the wind blew very fresh, we were obliged to wear her, and run back for the Main Tickle, through Friday's Bay, where we anchored about one in the morning.

October 6th.—At daylight found ourselves in a wide rocky channel, with bare and rugged cliffs on each side of us, and the wind blowing very fresh from the south-west. Passed through Herring Neck, and then tried to beat up for Dildo Run, hoping to get through into Exploits Bay in that direction. On getting from under the lee of the land, however, the cutter again refused to stay; and as it was now blowing hard with a tumbling sea, we bore away for Change Island Tickle, where we anchored. We then took in more ballast, and altered the trim of the little craft, to render her manageable if possible. There were many schooners here taking in fish for St. John's and other places. Several vessels had returned from the Labradore, giving very poor accounts of the fishery there. Some vessels had not caught fish enough for their own consumption, and none had completed a cargo. For the last eight years the fish seem to have been gradually deserting that shore. They are frequently very capricious in the choice of their haunts, and there seems to be a kind of periodical change taking place in this respect. Mackerel used formerly to be abundant in Newfoundland; but for the last few years not one had been seen, though it was believed they would eventually return.

October 7th.—As soon as it was light enough to see, I walked over the north island and shot a brace of ptarmigan; then weighed anchor, and we sailed along the east side of the Change Islands, and passing by an intricate channel through a number of low rocky islands surrounded by ioals and reefs, we made for

Gander Bay. It came on thick with heavy rain in the afternoon, so we anchored in shoal water on the west shore of the Bay.

October 8th.—Fine morning, with a light air from west-south-west. Sailed up Gander Bay, a long inlet with rather low and level shores, and anchored near its head in shoal water. The extraordinary shoals of Medusæ (called in Newfoundland squid-squalls) met with in these bays are most remarkable. Many of them were apparently turned inside out, and seemed to have fringes of eggs adhering to their internal filaments. We found one or two houses at the head of Gander Bay occupied by salmon-fishers. I had intended, if possible, to get into the large pond, but the only man at home here said it was now quite impracticable, from the small quantity of water in the brook. Under the most favourable circumstances it takes three days to get a punt up to the pond, although it only requires one to come down, the river being, as usual, full of rapids.* The rock about the head of the bay was a fine-grained dark slate. As nothing more was to be done without more

* A description of the pond and river, as derived from this man's information, will be found in the Report.

time than I had to spare, we returned.   As we sailed down the bay there occurred the best instance of sea mirage I had ever seen.   The low land on each side of the bay seemed to rise into the air in the distance and terminate abruptly in a point.   A boat sailing up the bay had the image of her sails inverted below her, the reflection being as bright and distinct as the real object ; and as we approached her, she seemed to rise preternaturally fast out of the water, her hull and its reflection being both simultaneously enlarged.   On arriving at the mouth of Gander Bay, it fell calm, and I took Simon and Tom in the boat after some birds, leaving the cutter in charge of the old skipper and a boy.   Unfortunately, Simon had been distributing the afternoon's grog, and had left a jug of rum on deck. This was a temptation old Jack Litton, the skipper, could not withstand ; and I suppose he made himself comfortable with it.   On coming on board again, at dusk, I went below to get some tea, and afterwards came on deck to see where we were going, as I heard, by the ripple, we were moving through the water. The moon had risen through some hazy clouds, and on looking about I could not make out our posi-

tion at all. I asked old Jack where we were
going, and he said we were entering Dog Bay,
but on comparing the chart and the compass,
I could not understand it. As he appeared
perfectly steady, and I had no knowledge of
the rum part of the business, I left him to manage his own craft for a little while, but on
looking about again, and finding he had altered
our course, I asked him where we were going
now, and he said we were going for Loo Bay.
By this time I perceived how matters stood,
and notwithstanding that he stoutly affirmed
he knew where he was, and pointed to Loo Bay,
Farmer's Arm, Stag Harbour, Tickle, and
Rocky Bay (places, some of them, thirty miles
off), I insisted on his walking forward and giving me the helm. As he was the only one of us
who had ever been in the neighbourhood before,
I saw my men and the boy were more inclined
to trust to his knowledge than to my charts,
and indeed I did not know exactly whereabouts
in the chart we were. However, sending Tom
aloft to look out, I kept on the course we
had been going, till we got close to a rocky
shore, on which in a quarter of an hour we
should have been wrecked, had we remained

under the guidance of old John. I then made out the opening between two of the islands, and thus got a point to start from, and by keeping Tom at the mast-head, and Simon at the helm, and diligently attending to the chart, I managed to pick our way through. The passage was rather intricate, the light of the moon dim and uncertain, and the headlands and islands appeared all so grouped and massed together, and most of them were so much alike, that it was by no means an easy task. I had constantly to run up and down all night long, to examine the chart, and compare it with the land about, taking care never to lose sight of one known point, before establishing another, and keeping account of all the bearings and distances. The breeze, too, was freshening, and to a person "not to the manner born" the mere necessity of rapidly translating the true bearings into compass bearings, for the guidance of the helmsman, was perplexing enough. Old John kept swearing we were all wrong, and giving all kinds of opposite directions and advice, but as the men ceased to pay any attention to him, we succeeded at last in getting into tolerable sea-room, and beat up for the mouth of Dildo Run. About four in the morning we anchored in a small cove, between Farewell Head and the Run.

October 9th. — We left this cove, called Beaver Cove, and ran into the mouth of Dildo Run this morning, but as the wind still blew from the south-west, we could not pass through it. Wishing to see as much as I could of the singular heap of islets and rocks in this channel, through the multitudes of which there is only one narrow passage of deep water in some places not more than thirty yards across, I landed on New World Island, and crossed through some woods to a marsh beyond which was a hill. It came on to rain most tremendously; so we got thoroughly wet, miserably cold, and had a hard climb for nothing.

October 10th. — A schooner came down through the Run this morning, in company with which we sailed for Change Island, as I began to be fearful lest all the St. John's vessels should sail and leave me behind. Under the lee of the high land of New World Island we had only a pleasant breeze, the wind being about north. As we proceeded, however, it blew harder, and

on getting down to Indian Garden we met a
fleet of boats, large and small, with mainsails
down, foresails reefed, and every sign of a heavy
breeze blowing outside. There were twenty-four
of them altogether, and they were running up
among the islands, some for shelter, and others
to go to their winter habitations. We still held
on, hoping to fetch Change Island Tickle, but
on getting exposed to the full fury of the gale,
our cutter, which was light and crank, heeled
over, till the water boiled over the rail, up nearly
to the foot of the mast. A sea washed over us,
and put out the fire, capsized the coffee-pot,
spoiled our breakfast, and I could hear all the
things in the cabin going to leeward in one
great crash. This would never do, and as we
could not hope to fetch the harbour if we took in
or reefed sail, we wore her round and ran back
again. The schooner followed our example,
and about two in the afternoon we anchored
under the lee of the land, having to beat up for
an hour or two before we could get sufficiently
close inshore. We then landed, made a great
fire, and had some breakfast. As long as the
wind was south of west, the weather had been
warm and pleasant: with this northerly breeze,

however, the salt water froze on our decks as it
washed over them, and the thermometer, at
noon, in the shelter of the woods, was scarcely
above the freezing-point. As the woods around
us were very thick, and there was no beach, we
could not walk about, and were reduced to
lying by the fire all the afternoon, warming
each side alternately, while the other froze.

October 11th.—Sailed with a light breeze
from the west, which soon carried us down to
Change Island Tickle, through a heavy sea,
tumbling and dashing over the rocks and
islands around us. On anchoring, we learnt
that all the vessels bound for St. John's had left,
but that one belonging to Mr. Slade was still
waiting in Fogo Harbour, to take Mr. S. and
Mr. Peyton, who were expected down to go on
board of her to St. John's. As I thought it
possible they might go down to Fogo outside
the island, I wished to go on. The people said
it was no time for any loaded craft to try to
get into Fogo Harbour, but our light cutter
might do it; and old John said he thought he
could take her in, so we set off. We found a
tremendous sea rolling in from the north, with
only a light wind blowing from the west,

making it more difficult to manage. Mountains
of foam were bursting over all the shoal rocks
about; Black Rock was alternately dry, standing thirty feet out of the water, and buried in
boiling surges that were raging and swelling
above its head. The scenery was very grand;
the bare dark rocks of Fogo Head frowning cha-
racteristically above the waves. On nearing the
entrance of the harbour we found the western
entrance under Fogo Head white with foam all
across, and as the wind would be likely to baffle
us under the high land, we prepared to run in
by the Boatswain's Tickle, a narrow passage
between two of the islands, of only two fathoms'
depth at low-water. John gave the men directions to stand by the main sheet, and when
he told them to haul, to pull away without
minding what became of themselves, and on
no account to let go. As I could be of no service, I held on by the rigging, amidships, to
watch the result. As we slowly drew in between the islands, the rocks on each side of us
presented themselves alternately bare and black,
and covered with torrents of white foam, in
which, as it extended across the passage, we were
soon involved. Then came a long thundering

wave which lifted us up on high, and launched
us into the middle of the channel.   Down we
sank as it retired, till we seemed touching the
rocks below, and our rudder did receive a slight
shock.   All our heads were then turned aft to
watch the approach of the next wave.   On it
came, a huge mountain of water, rising continually higher as it approached, lifting up our
punt, which was towed behind, high over our
heads, so that I verily thought it would fall over,
on to our deck, until at last it plunged under our
stern, and up we rose on the swelling wave,
which buried Tom and Simon for a brief space,
and drenched our decks as far as the foot of the
mast.   This wave hurried us helter-skelter close
to a black rock, upon which I could have leaped
ashore from the leeward rigging, as the water
receded, and to avoid which it had been necessary to haul in the main sheet.   Luckily, the
men obeyed their directions, and old John, as
steady as a rock when he had no rum on board,
never blenched from the helm, so we crept past
it ; and another wave, of rather less violence
than the preceding, launched us into smooth
water within the harbour.   Some people had
collected on the opposite cliff to watch our en-

trance, as it was certainly " a very close shave,"
and old John acknowledged afterwards he
would not have tried it had he known the sea
was so high. As we were on the slope of the
middle wave, and I looked over the bowsprit
and saw it pointing directly down to the rocks
at the bottom of the water, I really had expected it to catch on one of them, and that the
little vessel would enter the harbour by a kind
of summerset. As it was, however, in about
twenty minutes I was comfortably seated in
Mr. R.'s parlour, feeding on fish with not the
less appetite that I had narrowly escaped becoming their food. Many vessels had been
lost at this place. A large brig belonging to
one of Messrs. Slade and Co.'s establishments
was once seen off the harbour in a gale of wind,
but as it was supposed she would make for
Change Island Tickle, she was not watched.
The next evening, when the sea had somewhat
subsided, the attention of some one was caught
by a flag flying out of the water just on the outside of a rock in the western entrance of the
harbour, and on going down to look at it, he
found it was the flag on the top-gallant mast-
head of the brig, which had struck on the rocks

below, and sunk with all hands on board. I was
sorry to find also on arriving at Fogo, that, only
a few nights before, a vessel belonging to Messrs.
Slade, with a cargo worth 4000*l.*, had been
wrecked at the Wadhams. The captain, being
well acquainted with the coast, had made Cape
Freels in the afternoon, but, by some mistake
having taken the wrong passage through the
Wadham Islands, struck on a rock and knocked
a hole in the bottom of his vessel, which sunk,
barely giving the crew time to get out the boat,
and save themselves. The captain saved the
letters, which, with the clothes he had on, were
the only things preserved from the wreck.

October 12th.—Mr. Peyton and his son came
to-day, without Mr. Slade, who could not accompany them. They had arrived at Change
Island Tickle soon after we left it yesterday,
but Peyton said he knew better than to attempt
to get into Fogo till the sea had gone down.

October 13th.—We sailed in the "Content,"
a miserable little schooner of about forty tons.
She was very deep, being loaded with a cargo of
second-rate fish, that stunk most horribly; and
besides this she carried a deck-load of tierces
of salmon. We did not discover until we had

sailed that the forecastle also was full of fish up to
the beams, and all battened down, and that we
were to share the cabin with the crew.   There
were Mr. Peyton, his son, myself, and my two
men ; the skipper, who was also owner of
the schooner, and a crew of five hands,—in all
eleven.   The cabin was a small triangular
apartment, boarded off from the hold, having
lockers all round to serve as sleeping-places,
with a bench against the sides of the lockers,
and just exactly room on the floor for one
seaman's chest or long box which stood there.
The lockers were covered with filthy blankets, and the place had never been washed
since the vessel was built : a small fireplace,
with a little brick chimney, the top of which
was level with the deck, and two or three strips
of wood nailed against the bulk-head to faci-
litate our exit by the small aperture which
served for the companion, completed the conveniences of this execrable hole, the very efflu-
vium from which was enough to poison one not
accustomed to it.   The place would not hold
eleven persons by any system of packing, but
as there was always a watch of one or two on
deck, the rest managed to crowd in.   Miser-

able as was existence in such a vessel, there was only the alternative of sailing in it, or remaining where I was till the following June. Soon after we left Fogo Island a gale of wind arose from the west-north-west, with a heavy sea. Our decks were only about a foot and a half out of the water while in the harbour, and being so heavily laden, the vessel stood firm and upright without yielding either to the wind or the water: every wave, accordingly, washed clean across our decks, and she looked more like a low rock than a vessel. It was a mystery to me how she managed to swim at all. We reached, however, Cape Bonavista, on the evening of the 14th, and had a temporary lull under the land between it and Catalina. We seized the opportunity to make tea, and eat some biscuit, this being the first refreshment we were able to get that day. Soon after we opened Trinity Bay, and got into another tumbling sea, which reduced us to the alternative of being drenched on deck or stifled in the blankets below.

October 15th.—The wind headed us in the night, changing to the south, and obliging us to make for Trinity Harbour. We did but just manage to scrape round the Horse Chops, and sailed slowly along this lofty shore, with its perpendicular cliffs, at whose foot the waves were boiling and leaping, dead to lee-ward, and not a quarter of a mile distant. Peyton seemed rather astonished when I called him on deck to look at them; for the parting of a single rope would have sent us where we should not have had a single chance for our lives. It was, however, a grand sight. As the shore receded, we managed to crawl off it, and at length ran into Trinity Harbour. Here we saw the judge's vessel at anchor, and, having washed and dressed on deck, were soon on board her enjoying a good breakfast. We were most kindly received by the judge and all on board, as well as by Mr. D., the agent for Brooking and Garland, and were hospi-tably entertained during our stay in Trinity. We tried again to reach St. John's on the 16th, but after getting half across Trinity Bay were sent back again by a south-east wind and a fog. On the 17th we did not attempt it, but dined and slept on board the judge's vessel, who wished us much to stop till the

court was over, when he would take us on
to Harbour Grace. As we were both, however,
anxious to get to St. John's, we set off before
daylight on the 18th, with a fine breeze from
the north-west. We lay fasting in our sa-
voury berths until about one o'clock, when
having got round Cape St. Francis, we were
lying alongshore for St. John's. I was amus-
ing Peyton by considering what an excellent
dinner we should eat when we got on shore,
and what it should consist of, when I felt a
slight shock and heard a report, and on jumping on deck I was informed, with many blank
looks, that they feared we had sprung the mainmast. We immediately doused the mainsail,
and then consulted what we were to do. If the
wind shifted ever so little to the west, we should
not fetch the mouth of the harbour under the
foresail alone, and as a north-north-west wind
blows out of the harbour, or at least baffles in
the narrows, we feared we should not get into
the harbour, if we did fetch the narrows. If
we stood round Cape Spear, the coast trended
so much to the south-west, that we feared we
should never be able to fetch the Bay of Bulls,

or any other harbour, without the help of the
mainsail.   There was no other land in that
direction, and on inquiry we found there were
just two gallons of water on board, so standing
out to sea would not do : I proposed we should
run up Freshwater Bay and beach her, but
that involved the loss of the vessel and cargo,
and perhaps ourselves too, as the surf would
be heavy.   At last we determined to clap on
all the stays, guys, and tackling we could to
keep up the mainmast, and have a try at the
harbour, hoping at least to get into anchorage
and hold on.   Accordingly we kept as close
in to the rocks as we could, and on opening
the narrows we ran up the mainsail, and
though the mast gave a most ominous creak
on every tack, yet by making short tacks and
easing it as much as possible, we managed
fairly to beat into the harbour, and anchored
at a wharf-head.   When we landed our appearance must have been somewhat remarkable, and not the less striking as the day happened to be Sunday.   I had taken one summer and one winter pair of trousers with
me, but was now obliged to wear them both,

not merely on account of the cold, but also
that, as the holes in each did not exactly cor-
respond, I might get a mean or average pair
of trousers from the two. My jackets were
in a similar condition, while our faces were
grimed with smoke and dirt, and our whole
persons redolent of bad salt-fish. On examining the mast next day, it was found to be
broken short off just below the deck.

Thus ended my last excursion in Newfoundland, and I can only give my advice to any
one who wishes to lead the life of a traveller
to commence with this country, in order to get
well accustomed to rough living, rough fare,
and rough travelling, and to get rid of all delicate and fastidious notions of comfort, convenience, and accommodation he may have ac-
quired by journeying in England. I must
add, however, that so far as the inhabitants are
concerned, under a rough exterior he will meet
with sterling kindness and hospitality.

In November I sailed from St. John's in Her
Majesty's steam-ship Spitfire, which I men-
tion as she was the first steamer ever seen in a
Newfoundland port. She happened to touch

here in order to bring a few troops from Halifax, and great was the wonder and admiration she caused among the population of St. John's. Some boats and schooners outside were so astonished as she approached, that they had scarcely presence of mind to get out of her way, and she had very nearly run them down.

NOTE

NATURAL HISTORY

OF

NEWFOUNDLAND.

PREVIOUSLY to Professor Stuwitz visiting the country, I was preparing to make some inquiries concerning the Zoology of Newfoundland, but, on finding that the results of his voyage will be published by the Swedish and Norwegian government, I left the investigation entirely to him. In the mean while I here give a short account of those animals which I saw or heard of while in the island.

MAMMALIA.

The black bear is found in the wilder parts of the country, but is now becoming scarce. This animal is not savage except when wounded, and Mr. Peyton of Toulinguet informed me that a bear will run from a man, if he either sees or even scents him down the wind, just

as a deer will. In the summer and autumn
they live chiefly on berries, but I heard several stories of their coming out in the winter
to lone houses in the woods in search of food,
and of their being killed with hatchets. They
are, apparently, partial to pork and molasses.

The white or polar bear has been known
to land from the ice in Newfoundland, and an
instance has occurred of one having been killed
near St. John's.

The wolf has become more abundant of
late years. It is a tall, large, and very powerful animal, grey on the back and yellow beneath. Wolves have been shot in the outskirts
of St. John's during the winter, as well as at
many other places. They rarely, if ever, attack men, or even children, but are not so
timid as the black bear, as they will lurk about,
and dodge the steps of a traveller in the
country, in order to take advantage of any
accident that may happen to him. George
Lane of Freshwater Bay told me that he
walked up Gambo Pond on the snow one
winter's evening to visit a person living at
the head of it, and on his return the next day
found the tracks of two wolves, one on each

side of his own footmarks, who seemed to have regularly hunted him, as the tracks every now and then separated for a hundred yards or so, and then at regular intervals closed in again on his track and again separated. They appeared to have followed him one on each side in order to come on his track should he diverge, while they met occasionally to be sure that they had not passed him. From what I heard they seem to be fond of dog's flesh, and a single wolf appears to be more than a match for any Newfoundland dog. They must destroy a vast number of deer, and in the winter they do so much damage to the cattle, that a reward of 5*l.* for the head of every wolf is now offered by the Colonial Government.

The Caribons, or reindeer, are the only animals of the deer kind in Newfoundland. They are very large fine animals, with immense horns. How they manage to get through the woods with these horns has often puzzled me. I was told that they elevate the nose, lean their antlers on their shoulders, and then dash into the wood pell-mell. They always, however, choose the thinnest and most open parts, and a

deer-path is always found to be the best guide
from one marsh to another through the intervening skirts of wood. These deer-paths are
very numerous all over the country, and exactly
resemble sheep-walks; but besides the regular deer-paths, every marsh in the unfrequented
parts of the country is covered with the impres-
sion of deer's feet. These foot-marks are something like those of a cow, but wider and larger,
and they remain a long while visible on the
moss. Notwithstanding the abundance of deer
in many places I visited, I was never lucky
enough to see one alive, nor have I seen more
than one recently killed. They are, however,
often brought dead into St. John's during the
winter, where they sell at 15s. a quarter. The
venison, though not fat, is soft, juicy, tender,
and, in my opinion, most excellent meat. During the early part of the summer, they appear
to separate into pairs and to hide themselves in
the recesses of the woods. When the doe has
brought forth her young one, she is said to
conceal herself with it from the buck ; but this
is not always the case, as my men saw near
Chapel Arm, Trinity Bay, the buck, doe, and
fawn together, and I saw their traces myself.

This was in the latter end of June. The only
opportunity of seeing them in the summer is
when they are driven out of the woods by the
flies and the heat of the weather, and come to
bathe themselves in the lakes.

In September and October they are in the
best condition, and the buck is then fat. About
this time they migrate from the north towards
the south, swimming in herds across the lakes
and arms of the sea. During the winter they
are abundant on the south coast, and scarce
about Bonavista Bay and the north. Formerly
the herds that came down to the south coast
were immense. Mr. Bagg of La Froile told
me he had seen the shores of the bays about
there, before they were inhabited, literally
crowded with deer in the winter. He spoke of
many thousands, and assured me he once killed
seven at one shot with heavy slugs from a large
sealing gun. I heard also of two or three, and
in one case of five, having been killed at a shot.
In the spring, about March, they re-migrate
towards the north; and it is in their spring
and autumn migrations that the greatest number are killed by the Indians and others. They
shed their horns to a late period of their

lives, as the one shot by Sulleon on the Grand
Pond in the beginning of September, which
was an old buck, had the velvety covering
hanging in rags about his horns, one of which
was not fully grown. The colour of their skins
is said to vary slightly according to the season,
being darker in the summer than in the winter. They would probably be domesticated
very easily, as several fawns have been tamed
so far as to walk about the woods and return
to the houses, but they have in most cases
been eventually killed by the dogs. Were
the attempt made on a larger scale, and the
dogs taught to respect the deer, there is no
doubt of the domestication being possible, and
I think it might be made a profitable speculation.

The fox is tolerably abundant in Newfoundland; and there are, besides the common yellow or reddish fox, the varieties known as the
black, the silver-haired, the blue, and the white
foxes. The black, and silver-haired, are much
valued for their fur.

The hare also is found in some parts of the
island in considerable abundance. It is as
large as the English hare, but becomes of a

dirty white colour in the winter. Mr. Gray,
of the British Museum, informs me that he
believes it to be of a different species, or at
least a well-marked variety from the English
hare.

Martens were formerly abundant, but they
are now growing scarce; and the same is the
case with the beaver, which is now confined to
the most retired and inaccessible parts of the
interior. I saw but one, and that was skinned
and ready dressed for the spit, just as I was
leaving Fogo. They are considered by the
planters as excellent eating; but others have
told me they were too rich, and tasted too
much of spruce and other wood.

The musk-rat, or musquash of the Indians,
is very abundant, living in holes in the banks
of ponds, and about the mouths of brooks.
The otter is still very abundant, being found
principally in rocky brooks, but sometimes, as
in Random Sound, on the margin of the inlets
and arms of the sea.

Of the seals I have already spoken at large.
The only one, however, which can be said to be
an inhabitant of Newfoundland, is the Bay seal,
of whose habits some account has been given in

Vol. I. The others are merely occasional visit-
ants to the coasts for the purpose of bringing
forth their young, their native habitat being
farther north, on the coasts of Labradore and
Greenland.

Of the lowest order of mammiferous animals, the cetacea, there are abundance on the
coasts, including several species of porpoise,
grampus, and whale, but I cannot add anything to the incidental notices already given.

BIRDS.

There is at least one large species of hawk
or eagle, namely, the bird called the gripe,
which has a white head, brownish body, and
yellow legs: whether the one spoken of as
the fish-hawk be a different species or not I
cannot say with certainty, but should think it
must be.

The large snowy owl is abundant in Newfoundland; and there is also a brown owl,
which is not uncommon.

The common ptarmigan, of a reddish-brown
colour, with red about the eyes, and a few
feathers more or less white in summer, and
sometimes entirely so in winter, is the only

gallinaceous bird. It is feathered and haired down the legs, and between the toes.* These birds are universally diffused over the island, being much more abundant, however, in some spots than others. In clear weather, they are found about the skirts of the woods, and in the tucking bushes, and are then very wild and difficult to be got at, flying far out of sight when put up. When it is foggy, however, they come out on the barrens and marshes, and are very tame, merely flying a few yards, even when shot at, before they alight again. They usually squat very close, and the people who go to shoot them in foggy weather never take a dog, but stoop and look along the surface of the ground, and thus detect the head of the cock peering out from among the brown moss and bushes. The bird is distinguished at a considerable distance, by the red about the eye, and, getting within shot, the people sweep down a whole covey, firing at the birds on the ground in a most barbarous and un-sportsmanlike manner.

* I found a nest of this bird, when at Colinet, in June, with eleven eggs, of a brown or ash colour, spotted and mottled with black.

The bittern is not uncommon in Newfoundland : I saw two, but did not succeed in shooting them.

The wild goose I have already spoken of. It comes in May, breeds on the most retired and secluded ponds it can find, and brings its young ones down the brooks at the latter end of June and the beginning of July. In September the young are full grown, and at this time they all take their departure.

The black duck is a wild duck, only found in fresh-water during the summer. Its plumagè is a dark sombre brown : it is shy, but abundant in some spots ; and is a most delicate and excellent bird for the table.

The loo, or great northern diver, is also abundant. There are one or two species of large kingfishers : one or two of dark jays, one of which, of a dull grey colour, with very loose feathers, is almost the only bird that is ever seen in the woods. There are also several other smaller birds, among which the so-called blackbird, which is in fact a thrush, with a red breast and brown back, is the most conspicuous.

As summer residents, there are, besides those already mentioned, the snipe, in great abundance; plovers of two or three kinds; the curlew, in August; several small birds, called beach-birds, from their running along the beach either of seas or lakes; and probably many others with which I am not acquainted.

In the winter, a small bird, called the snowbird, comes in large flocks.

The marine birds are still more abundant than those of the land.

The gulls breed on the islands in the inland ponds. They are a great nuisance to the hunter, as they keep sailing over the woods at his approach, and alarming by their sharp cry every other animal about.

Eider-ducks, sheldrakes, cormorants called "shags," puffins, a kind of auk called "sea-pigeon," and several species of petrel, with many other birds, abound at various parts of the coast. One kind of sea-bird, abundant on the south coast in the summer, is called the "hound," from its cry. These "hounds" fly in flocks, and make a noise exactly like a pack of fox-hounds.

The beautiful white egret, with elegant drooping plumes on the back, and a very long thin neck, is shot sometimes, but is considered rare.

In hot summers, humming-birds have been known to visit the southern shores of Newfoundland. Mr. G——, the present sheriff of the island, told me, that once while he was accompanying the judge, and sailing alongshore near Fortune Bay, one or two humming-birds flew on board, which he caught, and took into his cabin, where he shut them up. Presently, one or two were again observed on deck, which were likewise caught and conveyed to his cabin. This occurred several times, till at last he thought he must have at least twenty shut up. At length, his suspicions were aroused by the same number always re-appearing, and on going to his cabin he found it quite empty, with a small window open at one end to let in air. The first birds caught had always taken advantage of this to escape to the deck again, where they had been as often re-captured.

REPTILES.

Of this class of animals Newfoundland (the Ireland of America) is entirely destitute.

Not a frog, nor a toad, nor a lizard, nor a snake, has ever been seen in the country. St. Patrick must certainly have visited it, in spirit at least, and effectually

"Banish'd all the varmint."

Fish.

The abundance of this class of animals perfectly compensates for the absence of all reptiles. Salmon and trout are the only freshwater fish, no other having ever been seen. The salmon are confined to the principal brooks, but the trout are frequently found in great abundance in the various ponds. They are, however, capricious in this respect; some ponds are full of them, while others, of precisely similar character, are without any. The eggs of the fish have, probably, been disseminated by adhering to the feathers of birds or the skins of other animals, which would account for the partial nature of their distribution. There seem to be at least two species of trout, the red and the white.

Of marine fish, besides the cod, the capelin, the lance, and the herring, there are large halibut, several kinds of flat-fish, none of which are ever caught; the large-headed spiny fish called the sculpin, several other smaller fish, the dog-fish, and the shark.

CEPHALOPODA.

The only cephalopod which I saw was the common squid, apparently the same as the Sepia loligo. They are generally eight or ten inches long, come to the shores in myriads in the month of August, and are caught in nets or picked up on the shore by the fishermen for bait. A shoal of them is perceived at a distance by the number of little drops of water, like rain drops, which each shoots into the air as he darts backwards near the surface of the sea. This appears to be their usual method of progression, and the forcible ejection of the water seems to be the principal impulse by which they move. I heard several stories of enormous animals of the squid kind; and Judge Carter, of Ferryland (who lately died at a very advanced age), told me of an animal having been seen in his father's time and drifted dead into the harbour which was a gigantic squid, and when cut up filled one or two barrels. These stories of enormous cephalopodous animals are prevalent in many parts of the world, and would seem to have at least some foundation in fact.

Conchifera.

In common with all the rest of the eastern coast of North America, there would seem to be but few species of what are commonly called shell-fish on the coast of Newfoundland, although individuals are sufficiently abundant.

Of gasteropoda, or univalve shells, I have only seen occasional fragments of Buccinum, apparently B. undatum, or the common English whelk, a small species of Littorina, adhering plentifully to the rocks in some places, and a large Natica, a specimen of which I saw at St. Lawrence. Of Acephala, or bivalve shells, there is a greater variety.

Two species of Mytilus, one growing to a large size, are very abundant. The smaller did not appear to differ from the common English species, and is not unfrequently eaten.

Large Pectens, or scallops, are found plentifully in some of the more shallow and retired harbours where there is mud or sand. They are eaten, but are indigestible and rather dangerous food. On the sandy shores of St. George's Bay and Codroy the Mya arenaria is found in the greatest abundance : the animals of this species are called " cocks and hens " by the fishermen, who sometimes use them as bait for the cod.

Two or three small species of Tellina were found in the same situation.

A species of Glycimeris, called " the clam," I have never found except in the stomach of the cod, but it is well known to the fishermen.

One or two species of fresh-water mussel, or Unio, may be found in some of the ponds and brooks. These are also used by the fishermen as bait, and called likewise " clams " or " glams." No oyster is known on the coasts of Newfoundland, but I saw and ate some of a good quality in St. Pierre which had been brought from Prince Edward's Island. They were very different from any English species, being oblong and narrow in proportion to their length from the hinge to the mouth of the shell.

The great depth of the sea along the shores of the greater part of the country is obviously unfavourable to the existence of many of the conchiferous mollusca, and those which exist can only be seen by dredging for them, or by finding them in the stomachs of fish.

Professor Stuwitz procured several Terebratulæ, of what species I do not know, in Fortune Bay : one of these he got alive, and dissected, drew, and described the animal.

On the great banks, where for many miles the water has not a greater depth than from twenty to fifty fathoms, a greater variety of shells may be got by dredging than near shore.

Professor Stuwitz informed me that on the southern part of the bank, and also to a certain extent on the southern coasts of Newfoundland, the shells and animals had a more tropical character than would be supposed from the latitude, and from the existence of other animals of an Arctic character. He attributed this to the action of the Gulf-stream, not so much from its raising the temperature of the surrounding water, as from its current sweeping the animals farther north than they would otherwise reach. As there is also a current coming from the north to the northern extremity of the Great Bank, it is probable

there may be a curious mixture, in the beds
now in the course of formation there, of the
remains of animals whose natural habitats are
widely removed from their present locality and
still farther from each other.

CRUSTACEA.

A large species of lobster, very like, but not
quite identical with, our English species, is
found in great abundance in some of the sheltered bays. These lobsters are very good when
cooked; but I found that in some places, more
especially in St. Mary's Bay, the people were
strongly prejudiced against them.

One species of crab, and several of the
smaller crustacea, such as, for instance, those
commonly called sea-lice, are abundant along
all the shores.

INSECTS

are likewise abundant, and there are many
of various kinds. I did not pay any attention
to them myself, but Mr. St. John, of Harbour
Grace, showed me a collection he had made, in
which were several beautiful butterflies and
other insects.

Radiata.

The seas in the neighbourhood of Newfoundland afford numerous instances of animals of this class. Of the genus Echinus, one species, the green Echinus, is very abundant, and is used as food by many of the French.

The common star-fish is also extremely plentiful in the rocks and shoals, while the species called Medusa's-head, with its numerous arms and fingers, is frequently brought up from deep water.

Professor Stuwitz has discovered several new species of Holothuridæ and allied animals, some of which are probably referrible to new genera. He has also made several observations and discoveries with regard to their structure and habits.

Myriads of individuals referrible to the genus Medusa, Beroe, and others of the Acalephæ, swarm in the seas around Newfoundland, and spaces, many square miles in extent, are sometimes filled with them. Doubtless there are corresponding multitudes of those microscopic animalculæ which form the food of these animals, and the phosphorescence of the sea points to the same fact.

Among the Polyps I only noticed one species of Actinia, two or three sponges, and a Myriapora; but besides this latter coral, there must be others, as one or two coral reefs are met with in the neighbourhood of the island.

GENERAL REPORT

OF THE

GEOLOGICAL SURVEY

OF

NEWFOUNDLAND,

DURING

THE YEARS 1839 AND 1840.

GEOLOGICAL SURVEY,

&c.   &c.

---

1. INTRODUCTORY REMARKS.

THE object of a Geological Survey of any country is, to become acquainted with its solid structure; to know what are the different earthy or mineral masses of which the country is composed, and what are their relative and actual positions.

As a general term for designating all those masses of earthy or mineral matters which are extensive enough to be considered as forming an integral part of the structure of any country, whether the masses be hard or soft, the word "Rock" is used.

Rocks are of two kinds: stratified or aqueous rocks; and unstratified or igneous rocks.

The stratified or aqueous rocks consist of beds or strata resting one upon another in regular order, which beds vary in thickness from a few inches to several feet, and in extent from a few square yards to hundreds of square miles. The unstratified or igneous rocks have

no regular beds or strata, but occur in all
forms, from veins a few inches wide, to irre-
gularly-shaped mountain masses : they are
commonly found or may be traced beneath
the stratified rocks, but sometimes cut through
them like perpendicular walls (in which case
they are called dykes), pass among them in
irregular and often tortuous veins, or, finally,
rest upon them in great masses.

The aqueous rocks have been formed by the
deposition of earthy matters from water : the
igneous rocks have once been in a molten state,
from intense heat. Where the igneous rocks
join the aqueous, the latter are often much
changed in their structure and mineral character : they are then called metamorphic or
altered rocks.

The beds of the stratified rocks are rarely
horizontal, but are generally inclined at a
greater or less angle to the horizon. This
inclination is called the " dip." By reason of
this inclination successive beds rise one after
another to the surface of the ground, where
they may be measured and examined, and in
this way a knowledge may be gained of the
structure of a rock to an extent far beyond that

which could be attained by direct excavation. The edges, or "outcrop," of a bed, or set of beds, may, under favourable circumstances, be traced for long distances running through a country, and showing themselves here and there, where the soil or other superficial matters are removed. The line of bearing along which this "outcrop," or appearance of the beds at the surface may be traced, is called the "strike" of the beds.

Masses of igneous rock frequently form long ranges of hills, on the sides of which the stratified rocks repose, and from which they dip or incline downwards on either hand. The range or strike of the stratified rocks is generally persistent throughout their course; the dip, however, is often varied, and the beds are frequently undulating, or curved up and down into great ridges and furrows, the curves running along straight lines sometimes for many miles. The curves are of varied extent: they sometimes have a radius of only a few feet or yards, but are sometimes on a grander scale, the opposite sides of the curve being several miles apart. In all cases the lines along which the curves take place are called "anti-

clinal " and " synclinal " lines. The anticlinal
line is that which runs along the top of an
upward curve, along the crown of the ridge,
as it were, and *from* which the beds decline
on either hand; the synclinal line is that
which runs along the bottom of a downward
curve or trough, and *towards* which the beds
decline on either hand. In either case the
curvature of the beds is not always shown on
the surface of the ground, which may be quite
level, the beds having been broken, degraded,
or cut down, to a certain height.

As a consequence of the inclined position of
the beds, it will be seen that the lowest portion
of the series may frequently occupy the highest
situations above the level of the sea, rising up
on the slope of the hills from beneath all the
others; and that, conversely, as we travel from
the hills to the plains, we frequently come
upon higher strata, beneath which the beds
we have left behind have successively passed;
and thus the highest beds of the series may
occupy lands but little above the level of the
sea, the other strata lying beneath them in
those places.

The stratified rocks consist of an infinite

number of beds, composed chiefly of clay, sandstone, limestone, &c., but containing occasionally beds of coal, salt, gypsum, or other
matters.

Different portions of this great series of
stratified rocks, having certain characters in
common, are classed together, and spoken of in
the aggregate under the term " formations."
These common characters are of several kinds,
but it will be sufficient here to say that they do
not necessarily include identity of mineral structure. All the formations, as also all the beds
and subdivisions of which they are composed,
have a regular order of superposition, which
may be broken or incomplete, but is never re-
versed. In other words, we never find a bed, or
a formation, lying above another in one place,
and below it in any other locality. Both the
formations themselves and the beds of which
they are composed, however, thin out in various
directions, and gradually come to an end. Thus
by the gradual thinning out of some beds and
simultaneous thickening of others, a formation,
without losing its identity, or much altering
its general characters, may yet be composed
of an entirely different set of beds in one place

from what it is in another, and the mineral characters of the beds may either be the same or totally different. In the same way the series of formations, by the thinning out or absence of some, and the increased thickness of others, may vary in different countries, the total thickness of stratified rocks remaining the same. This thickness, however, is not always constant; and sometimes, by the absence of certain intermediate beds or formations, two portions of the series may be brought together in one locality which are widely separated in others. It is possible that these two formations may still be parallel to each other, that is, may both be horizontal or inclined at the same angle with the horizon. This, however, is not often the case with beds whose position in the general series is very different, or, in other words, which are of greatly different age. It is more usual to find the lower formation inclined at a considerable angle with the horizon, and the upper one resting upon it horizontally, or at a less angle with the horizon.

This discordance of position is called " unconformability," and when any two formations resting one upon the other differ either in the angle of their dip or the direction of their strike, they are said to be "unconformable" to each other. This unconformability is sometimes produced, not by the absence of any of the formations, but simply by a difference in their position. Thus, if, of eight groups or formations, the four lower ones are highly inclined, while the four higher are horizontal or nearly so, it is obvious that the upper ones may rest successively upon the up-turned edges of each of the lower ones.

As may be expected, from the fact of their highly inclined and often disturbed position, the stratified rocks are frequently traversed by cracks or fissures, running through the beds sometimes for many miles in a straight line, and extending from the surface to unknown depths. On the opposite sides of these cracks the beds are frequently found to be at very different levels, the difference varying from one yard to upwards of a thousand, one portion having sunk below or been raised above the other to that extent. These cracks are termed "faults." They vary in width from a few inches to several yards, and are sometimes

filled with broken portions of the adjacent
beds, sometimes with clay, sand, or other matters.   The surface of the ground frequently
gives no indication of the existence of these
faults, even where the movements of elevation
or depression have amounted to hundreds of
feet.   Faults are often met with unexpectedly
in mining operations; and, whether the geologist can detect it or not, the occurrence of a
fault in any particular locality must always be
looked on as an *accident*.

In certain formations, generally those of a
more ancient date, and in igneous as well as
aqueous rocks, are found " mineral veins."
The origin of these is yet a matter of some
controversy.   In whatever way they were produced, however, they are, when they occur in
the stratified rocks, frequently neither more
nor less than faults, containing, among other
things, minerals in a crystallized form.   Whether in igneous or aqueous rocks, they may
always be described as fissures traversing the
rocks, generally at a great angle with the horizon, uncertain in their occurrence, and as much
matters of accident as faults.   No rules can be
given for their discovery, except that they are

more frequent in the neighbourhood of granitic
or other igneous rocks, and where the stratified rocks have been greatly shattered and con-
vulsed. Metallic ores are sometimes found as
bunches, nests, or strings disseminated through
the rocks : these, however, are even more capricious and irregular than what are commonly
called mineral veins.

As a general fact, then, every large district of the earth is composed of igneous and
aqueous rocks, the latter resting on the former.
The aqueous rocks are composed of many
widely extended layers or strata of earth, resting one upon another, either in a regular series,
or in such a way as to be referrible to a regular series. In different places, these strata
are tilted up, broken, contorted, and cracked,
and are here and there pierced through by
igneous rocks, which, having been injected
among them in a molten or fluid state, occur
in every imaginable form and condition.

Such being a slight general sketch of some
of the characters, positions, and relations of the
materials of which the crust of the globe, or
at least that portion of it above the level of
the sea, is composed — it is evident that the

geological surveyor, in entering upon a country with which he has no previous acquaintance, has a task of some difficulty, and often
of great perplexity. He must make himself
acquainted with the characters of the various
beds constituting the series of stratified rocks
of which the country is formed. He must
discover the natural order of this series, and
ascertain what are the general characters
which are common to certain large portions
of it, in order to divide it into groups or formations. He has to accustom his eye to these
characters, in order to detect them where they
may be obscure, and many other little points
of detail must be worked out before he is
thoroughly qualified to enter on his task of
describing the solid geometry of the district.
For the more perfect fulfilment of this, too,
it is necessary that the country should be
accessible in every direction, that the surface
should not be too much covered up by vegetable soil, or hidden by a thick growth of
wood, and that the cliffs by the sea-shore, the
natural sections exhibited in the beds of rivers,
or the banks which enclose them, and the ap-
pearances on the hill-sides, or, in their ab-

sence, artificial cuttings, sinkings, or borings should be attainable. The very first requisite is, that a good map of the country should be placed in his hands, showing the general ranges and heights of the hills, and the courses of the rivers and valleys; and that this map should be on a sufficiently large scale to enable him to trace his daily routes and mark down the phenomena observed. Furnished with this knowledge and these materials, it is the business of the geological surveyor to trace on the map the surface-boundaries between the igneous and aqueous rocks, as also between the different formations or subdivisions of the latter. In doing this, he must also collect such information from the continued observation of the dip, strike, faults, veins, and other phenomena, as shall enable him to show the positions of the rocks below the surface, to as great a depth as he is able. This is done by means of sections, or supposed perpendicular cuttings through the earth in various directions, by which its internal structure may be best exposed. By means of these coloured maps and sections, a cabinet of specimens, and written descriptions, the geologist is enabled

to convey an intelligible idea of the physical
structure of the country, and the qualities and
properties of the materials of which it is composed.

The results of a geological survey, if fully
carried out, will be a knowledge of the existence and situation of all beds of good building-stone, lime-stone, slate, gypsum, coal,
iron-stone, salt, or other regularly bedded
rocks. It will probably point out the localities in which mineral veins will be most likely
to reward the search of the adventurer. It
will show the boundaries of the different districts best adapted for agriculture ; and will
be of the greatest use in draining, well-sink-
ing, road-making, and all other operations of
the farmer, the engineer, or the miner. Nor
must it be forgotten by the geological surveyor, that while he is bound by every means
in his power to discover and lay open the
natural resources of the country under examination, for what are commonly called
practical purposes, he must not neglect the
theoretical portion of the science of geology.
A knowledge of the mode of formation of
the rocks and the cause of their being placed

in their present positions, will constantly help him in difficult or obscure points; and will more especially be of the greatest possible assistance to him in mining or other operations, in which direct observation is impracticable. He is bound, therefore, to store up every fact, and even every speculation that may rise in his mind from the observation of fact, bearing on theoretical geology, and to contribute, as far as he is able, to the great mass of knowledge which is gradually working out for us the history of the formation of the present crust of the globe.

Those who are acquainted with the island of Newfoundland will be best able to appreciate the difficulty of applying the general principles and rules laid down in the foregoing observations to its particular case. The interior being trackless, uninhabited, and obscured by woods and morasses, the coast affords the only means of continued observation. Here, though the cliffs are bold, they are frequently inaccessible, and often either too perpendicular or too well guarded by surf to render landing practicable. Not only is there no map of the interior, but no general

knowledge of it exists.   No guide can be found
who knows more of the country than a few
miles round his own dwelling, or a particular path to a neighbouring settlement.   Much
time was therefore necessarily devoted to
gathering materials for a rough map and ac-
quiring some information on the physical geography of the country ; and the present Report can only be looked on as a collection of
so much preliminary information as would
have been of use in the commencing a detailed
survey, had the nature of the country rendered such an undertaking advisable.

Sketch of the Physical Geography of Newfoundland.

THE general character of the island of Newfoundland is that of a rugged, and, for the most part, a barren country. Hills and valleys continually succeed each other; the former never rising into mountains, and the latter rarely expanding into plains.

The hills are of various characters, forming sometimes long flat-topped ridges, and being occasionally round and isolated, having sharp peaks or craggy precipices: the valleys, also, vary from gently sloping depressions, to rugged and abrupt ravines; and the sea-cliffs are for the most part bold and lofty, with deep water close at their foot. Great boulders, or loose rocks scattered over the country, increase the general roughness of its appearance and character. This uneven surface is covered by three different kinds of vegetation, forming districts, to which the names of "woods," "marshes," and "barrens," are respectively assigned.

The woods occupy indifferently the sides or

even the summits of the hills, and the valleys,
and lower lands.   They are most commonly
found, however, clothing the sides of hills, or
the slopes of valleys, or wherever there is
natural drainage for the surplus water; and
perhaps for this reason they occur in greatest
abundance in the neighbourhood of the sea-
coast, or round the lakes and rivers, if the soil
and other circumstances be equally favourable.
The trees consist, for the most part, of fir,
spruce, birch, pine, and juniper, or larch;
and in certain districts the wych-hazel, the
mountain ash, the alder, the aspen, and some
others, are also found. *   The character of the
timber varies greatly, according to the nature
of the subsoil and the situation.   In some
parts, more especially where the woods have
been undisturbed by the axe, trees of fair girth
and height may be found: these, however, are
either scattered individuals, or occur only in
small groups.   Most of the wood is of small
and stunted growth, consisting chiefly of fir-
trees about twenty or thirty feet high, and not
more than three or four inches in diameter.

* One or two good-sized trees, apparently identical with
the common English ash, were seen near St. George's Bay.

These commonly grow so closely together, that their twigs and branches interlace from top to bottom, and lying indiscriminately amongst them, there are innumerable old and rotten stumps and branches, or newly fallen trees, which, with the young shoots and brushwood, form a tangled and often impenetrable thicket. The trees are often covered with lichens, and tufts of white dry moss are entangled about the branches. Other green and softer mosses spread over the ground, concealing alike the twisted roots of the standing trees, and the pointed stumps of those which have fallen, the sharp edges or slippery surfaces of the numerous rocks and boulders, and the holes and pit-falls between them. Every step through these woods is a matter of toil and anxiety, requiring constant vigilance to avoid falling, and constant labour to procure standing-room; climbing and creeping, and every mode of progression must be had recourse to, and new directions have constantly to be taken, in order to find the most practicable places through which to force a slow and tortuous way. In the heat of summer, while the woods are so thick as to shut out every breath of air,

they are at the same time too low and too
thinly leaved at top to exclude the burning
rays of the sun, the atmosphere being rendered close and stifling by the smell of the
turpentine which exudes from every pore of
the trees.

Embosomed in the woods, and covering the
valleys and lower lands, are found open tracts,
which are called "marshes." These marshes
are not necessarily low, or even level land, but
are frequently at a considerable height above
the sea, and have often an undulated surface.
They are open tracts, covered with moss to a
depth sometimes of several feet. This moss is
green, soft, and spongy, and is bound together by straggling grass, and various marsh
plants. The surface is very uneven, abound-
ing in little hillocks and holes, the tops of the
hillocks having often dry crisp moss like that
on the trees. A boulder or small crag of rock
occasionally protrudes, covered with red or
white lichens, and here and there is a bank
on which the moss has become dry and yellow.
The contrast of these colours with the dark
velvety green of the wet moss frequently gives
a peculiarly rich appearance to the marshes,

more especially when they are seen from a little distance, clothing the undulating slopes of a hill with tufts or thin skirts of wood scattered about. Except in long-continued droughts or hard frosts, these marshes are always wet, and unable to bear the weight of a person walking over them. A march of three miles, sinking at every step into the moss, sometimes knee deep, and always as far as the ancle, is, it may well be supposed, a most toilsome and fatiguing operation, especially when, as must always be the case in attempting to penetrate the country, a heavy load is carried on the shoulders. This thick coating of moss is precisely like a great sponge spread over the country, and becomes at the melting of the snow in the spring thoroughly saturated with water, which it long retains, and which every shower of rain continually renews. Numerous small holes and pools of water, and in the lower parts small sluggish brooks or gullies, are. also met with in these tracts, but it must be observed that the extreme wetness of the marshes is due almost entirely to the spongy nature of the moss, as the slope of the ground is nearly always suffi-

cient for surface-drainage ; and when the moss
is stripped off, dry gravel or bare rock is
generally found beneath.

The "barrens" of Newfoundland are those
districts which occupy the summits of the hills
and ridges, and other elevated and' exposed
tracts. They are covered with a thin and
scrubby vegetation, consisting of berry-bearing plants and dwarf bushes of various species,
and are somewhat similar in appearance to
the moorlands of the north of England, differing only in the kind of vegetation, and in
there being less of it. Bare patches of gravel
and boulders, and crumbling fragments of
rock, are frequently met with upon the barrens, and they are generally altogether destitute of vegetable soil. It is on the barrens
only, of any part of the interior of Newfoundland, that it is possible to walk with
any kind of ease or expedition : their hard
ground, though frequently broken, rugged, or
precipitous, being delightful to tread after traversing the heavy marshes or toiling through
the tangled woods. Sometimes, however, in
the hollows of the barrens, as also in other
situations, a bed of dwarf juniper is met

with, which goes in Newfoundland by the name of "tucking bushes." These grow about breast-high, with strong branches at right angles to the stem, and stiffly interlacing, their tops being flat and level, as if they had been mown off. They are so stiff, that in some places one can almost walk upon them; but as this is not quite possible, the labour of wading through them may be more easily conceived than described.

These different tracts are none of them of any great extent; woods, marshes, and barrens frequently alternating with each other in the course of a day's journey. In describing the general features of the country, one of the most remarkable must not be omitted, namely, the immense abundance of lakes of all sizes, all of which are indiscriminately called "ponds." These are found universally over the whole face of the country, not only in the valleys, but on the higher lands, and even in the hollows of the summits of the ridges and the very tops of the hills. They vary in size, from pools of fifty yards in diameter, to lakes upwards of thirty miles long and four or five miles across. The number of those

which exceed a couple of miles in extent
must on the whole amount to several hundreds, while those of a smaller size are absolutely countless. Taken in connection with
this remarkable abundance of lakes, the total
absence of anything which can be called a
navigable river is at first sight most ano-
malous. The broken and generally undulated
nature of the country is no doubt one cause
of the absence of large rivers. Each pond, or
small set of ponds, communicates with a
valley of its own, down which it sends an in-
significant brook, that pursues the nearest
course to the sea. The chief cause, however,
both of the vast abundance of ponds, and the
general scantiness of the brooks and smallness
of the extent of each system of drainage, is to
be found in the great coating of moss that is
spread over the country. On any great accession of moisture, either from rain or melted
snow, the chief portion is absorbed by this
large sponge, the remainder fills the numerous
ponds to the brink, while only some portion of
the latter runs off by the brooks. Great periodical floods, therefore, which would sweep
out and deepen the river channels, are almost

impossible, neither have the rivers power at any time to breach the barriers between them and unite their waters. In dry weather, when, from evaporation and drainage, the ponds begin to shrink, they are supplied by the slow and gradual drainage of the marshes, where the water has been kept as in a reservoir to be given off when required. In this way, many ponds, which, having no great depth, would otherwise be exhausted, are kept full of water in the driest seasons, and it is only in the greatest droughts, when the marshes themselves begin to dry up, that the ponds are found to shrink much below their average level. The quantity of ground covered by fresh water has been estimated, by those acquainted with the country, at one-third of the whole island, and this large proportion would not probably be found an exaggeration.

The province of Avalon is nearly separated from the rest of Newfoundland by the bays of Placentia and Trinity, a narrow isthmus (at one part not more than three or four miles across) alone intervening.

There are in Avalon two principal ranges of hills, forming regular watersheds.

The most easterly range is that which runs
from the back of Renews to Holyrood, in Conception Bay. Though not lofty, it is very
rugged, the faces of the hills being precipitous. At each end of this range is a remarkable hummocky hill called the " Butterpots ;" and so little is the country penetrated
or known, that it is a common belief among
the inhabitants of the neighbourhood, that the
Butterpots of Renews is the same hill with the
Butterpots of Holyrood, whence their common
name. They are distant about twenty miles
from each other, and on account of the distance,
and the intervening hills, neither of them can
be seen from the summit of the other. Each of
them appears to be about the same height above
the sea, probably rather more than 1000 feet ;*
and eminences of nearly equal altitude occur
at several parts of the range to which the local
names of Bold Face, Bread and Cheese, the
Drop, the Flakey Downs, &c., have been at-
tached. One considerable hill detached from
the general range is seen a few miles north of

* I was so unfortunate as to break a mountain-barometer
during the ascent of one of these hills, and therefore could
not ascertain their exact altitude.

Cape Broyle Harbour, and is called Hell Hill.
From the Butterpots of Holyrood a range of
high land runs along the eastern side of Conception Bay, forming the White Hills, and the
lofty iron-bound coast, from Topsail Head, by
Portugal Cove, to the neighbourhood of Cape
St. Francis. Another spur of this range, also,
runs towards the western side of Conception
Bay, as far as the Cat's Cove Hills, near Collier's Bay, these latter hills being the three
peaks of a picturesque elevation, and rising to
a height of 900 feet.

The other principal range of Avalon runs
from Cape Dog in St. Mary's Bay, to the
neighbourhood of Chapel Arm in Trinity Bay
It is less broken and rugged than the one before mentioned, forming a more level and continuous ridge, on which are various elevations,
for the most part rounded and flat-topped.
The principal of these are called Mount Sea-
Pie, North Harbour Lookout, the South-east
Mountains, Cap Hill, the North-east Mountain, the North Hill, Spread Eagle Peak, Little
Gut Lookout, the Tolt, and the Monument. Of
these hills, the north-eastern mountain of Placentia (distant about nine miles from the head

of the north-eastern arm of that harbour) is the most lofty and considerable. It is a round-topped hill of gentle ascent, and probably rises to the height of 1200 or 1400 feet above the sea. * From the top of the North-east Mountain sixty-seven ponds were counted, some of them two or three miles across, none less than 100 yards, and none at a greater distance than ten miles from the base of the hill. Many more existed within that area hidden by woods and intervening hills.

As subordinate hills and ranges, may be mentioned the Sawyers' Hills, with a peaked serrated outline a few miles south of Placentia; the ridge of high rough land forming the isthmus connecting Avalon with the main part of the island; Spaniard's Bay Lookout, and the high lands running down the peninsula between Trinity and Conception Bays; Chisel Hill, to the north-east of St. Mary's Bay; and lastly, the ridge of the South Side Hill,

* My barometers being broken, I could only judge' of its height by the distance of the sea horizon. The hill in question was seen some miles beyond Cape English in St. Mary's Bay, which is distant thirty miles in a straight line. Reckoning, then, the distance of the horizon at forty miles, the height of the hill would be about 1300 feet.

running from Torbay to the Bay of Bulls, Branscombe Hill, and the other elevations about St. John's.*

The only basin of drainage in Avalon worth mentioning is that lying between the two principal ranges of hills, the waters of which are poured into St. Mary's Bay. A number of ponds, however, lying south of Renews Butter-pots (some of the eighty which were counted from the summit of that hill), give rise to a considerable brook which empties itself into Biscay Bay, Trepassée. The rivers emptying themselves into St. Mary's Bay are, the Salmonier Brook, flowing out of a large pond called the Hundred Island Pond, and Colinet and Rocky rivers, emptying themselves close together into Colinet Arm. Of these, Rocky River is the most considerable, having the longest course of any river in Avalon. At its mouth it is 150 yards wide, and for half a mile is several feet deep, with rocky and precipitous banks. Its navigation is then stopped by waterfalls of very picturesque appearance: the river here,

* The principal heights about St. John's are Signal Hill, 520 feet; the Southside Hill, about 700; and Branscombe Hill, 870 feet above the sea-level.

forty yards across, takes two leaps of about
twenty or thirty feet each, over ledges of hard
rock, with a foaming rapid of 100 yards in
length between the falls.   Above this spot the
river is from sixty to eighty yards in breadth
for some miles, but rapid, stony, and rarely
more than knee-deep.   About twelve miles up
it forks, and the principal branch, called Hodge
River, takes its rise from some ponds which
are not more than five or six miles distant from
Brigus in Conception Bay.

To enter on the description, at greater detail,
of the many smaller hills and brooks of Avalon
would be waste of time, more especially as the
more remarkable of them are marked on the
rough map which accompanies the Report.

Of the physical geography of the main portion of the island of Newfoundland I can only
give a very slight sketch, taken from two or
three unconnected points.

The western shores of Placentia Bay and its
adjacent islands, from Cape Chapeau Rouge to
Piper's Hole, are rugged and precipitous, with
every character of a mountainous country in
miniature, none of the hills exceeding 1000
feet in height.   Cape Chapeau Rouge and the

neighbouring hills of St. Margaret and St.
Anne, probably attain a height of 800 feet, and
hills of equal elevation are seen occasionally
along the coast to the northward. The narrow
island of Merasheen cannot be less at some
points than 600 feet high. This same range
of lofty, broken, and precipitous land runs
along the western side of Trinity Bay, down to
Trinity Harbour, and thence crosses into Bonavista Bay about Keel's Head. It has an
irregular width of several miles, occupying the
eastern half of the peninsula between Fortune
and Placentia Bays, and forming a fine peaked
and serrated mass of hills some miles west of
Random Sound in Trinity Bay, which stretch
also to the neighbourhood of Goose Bay in
Bonavista. It is probably to the same range
that we must refer Gerard's Hill, Mount Stanford, the Lonil Hills, and the other high and
precipitous lands about the many islands and
sounds of the southern part of Bonavista Bay.
One remarkable hill at the head of Trinity
Bay deserves mention for the extensive view
it commands. It is called, in Trinity Bay,
Sainters Hill; on the chart, Centre Hill; but is
known in Placentia Bay by the name of Pow--

der-horn Hill. It is an isolated peak, upwards of 1000 feet above the sea, and overlooks nearly the whole of the Bays of Placentia and Trinity, as well as some of the high grounds about Conception, Bonavista, and Fortune Bays. From the summit of this hill 153 ponds were counted of all sizes, some of them having a diameter of two miles, and all within a radius of five miles on one side and ten miles on the other. It is evident that to map such a district as this with any degree of accuracy, and within any reasonable time, would be a hopeless task.

The western side of Bonavista Bay, from Clode Sound northwards, is comparatively low. As far as could be seen from the top of the Lonil Hills, or elevations farther in the interior, the country consists, towards the west, of regularly undulating ridges, running generally about north-north-east and south-south-west, never rising to a greater height than 300 or 400 feet, and covered for the most part with dense wood. At the northern extremity of Bonavista Bay, about Cape Freels, the country is still more low and level, and the woods are much thinner. For several miles on each side of Cape Freels there are fine level sand beaches, with shoal water, gradually deepening out to sea. The only other instances of such an occurrence are between Langley and Miquelon, and along the south side of St. George's Bay.

To the westward of Bonavista Bay, there are two ranges of hills, parts of which I saw at a distance, but neither of which was I able to visit. The first is that lying between the waters which empty themselves into Gander Bay, and those which fall into the Bay of Exploits. The northern end of this range is called the Blue Hills, and its continuation to the south goes by the name of the Heart Ridge. It runs about north-north-east and south-south-west, in a line with the promontory between Gander Bay and Dildo Run. No part of the range can exceed 1000 feet in height, as it is not visible from the coast.

The next range of hills crosses the River Exploits at about thirty miles from its mouth. The part to the north of the river gradually rises to the north into a summit called Hodges Hill; that to the south of the river goes by the name of Shutebrook Hills. These are seen from the mouth of the River Exploits, closing

the view up the valley of the lower part of
the river. They are flat-topped, with precipitous sides, having thus a square appearance,
whence the more southerly of them is called
Square Hill. Their height may probably be
about 1500 feet. A ridge of high land runs
from them towards the south-south-west, and
I was informed that another lofty hill had
been observed far in the interior nearly on
the same line of bearing.*

The southern portion of the island of Newfoundland, from the neighbourhood of Fortune
Bay to Cape Ray, is barren and desolate in the
extreme. About Cape La Hune the cliffs are
very lofty, and the high land close upon the
coast excludes all view of the interior from the
sea. West of the Burgeo Islands the cliffs are
low, the land rises gradually from the sea to
the interior, and the coast is fringed with
low islands, and rocks above and under water.
The country, as far as could be seen, is of the
most rugged and broken character possible,

* It is possible that this may be " Mount Misery," or else
" Jameson's Mountain " of Mr. Cormack. A very lofty
hill was mentioned to me by an Indian as visible at a distance of forty miles to the south-east from Orie's Hill near
the Grand Pond : this also is probably the same hill.

grooved in every direction by small valleys or
ravines, and covered with round hummocky
knobs and hills with rocky and precipitous
sides. This rocky country is partially covered
by moss, and low bushes, and berry-bearing
plants; but the only trees are a few dwarf
firs, huddled together in some more sheltered
nook, where a little soil has lodged that may
form a support for their roots. Nothing like
a prevailing direction or grouping together of
the hills is perceptible in this kind of country.
On approaching Cape Ray, however, a distinct ridge of a more decided character is perceived, running from the cape into the interior
in a north-east direction. Three remarkable
sugar-loaf hills rise from the low land which
forms the projecting point of the cape, but beyond them the hills are one unbroken ridge,
with a very steep face towards the north-west,
but rather flat and regular at top. This flat-
topped ridge, which here and there probably
attains a height of 1000 feet, is very well seen
from the Great Codroy River, forming the
southern boundary of the valley for twenty
miles at least; and from the top of the cliff near
Crabbs River, on the south side of St. George's

Bay, it was also seen running at a distance of about twenty or twenty-five miles towards the head of the bay. At the head of the bay, hills of various character and height are seen to sweep round from the south, and coming within eight or ten miles of the harbour, to stretch off towards the north. They here form not so much a connected chain of hills, as a band of broken and hilly country, frequently without any well-marked or definite boundary. One conspicuous hill bears true north-east from St. George's Harbour, distant about twenty miles in a straight line. It was called Hare Hill by an Indian, is a round bare-topped hill, and is within three or four miles of the south-west end of the Grand Pond. This range of hills runs from St. George's Bay northwards to the Bay of Islands, where it is cut through by the valley of the Humber River. It continues thence to the north, at least as far as Bonne Bay ; and from what I could learn of the country, as well as from the hills laid down on the chart, it appears to be continued thence down the centre and eastern side of the projecting tongue of land which stretches out to the north between White Bay and the straits of

Belle Isle. To the west of this chain of hills, which on the whole is the longest* and best marked in the island, lie two tracts of land, one on the south side of St. George's Bay, and the other between that bay and the Bay of Islands. The former is generally low and level, except the corner towards Cape Anguille, where it gradually rises to the height of 300 or 400 feet, forming the hills north of the valley of Codroy River. The latter tract has some low and generally level land to the west of Port-au-Port, but from Indian Head to York and Lark harbours there is a line of rugged and hilly country, forming the lofty peaked hills about those harbours, and the heights called " The Blow-me-down Hills," on the south side of Humber Sound. These attain the height of at least 800 feet, and in August, 1839, large patches of snow still rested in the hollows of their northern slope.

The principal lakes and rivers in the main part of the Island of Newfoundland are, the Grand Pond and Humber River, the Red Indian Pond and River Exploits, the Gander

* On this account I propose, in the present Report, to distinguish it by the name of the " Long Range."

Bay Pond and Brook, and the ponds called by
Mr. Cormack " George the Fourth's Lake,"
" Jameson's Lake," and " Bathurst Lake."

The Grand Pond is full fifty miles long,
and about five miles across at its widest part,
namely its north-eastern extremity.   Its south-
western extremity bears about north-north-
east from the head of St. George's Bay, and is
about fifteen miles distant from it.   It cuts
deeply into the eastern side of the long range
of hills before mentioned, and receives several
brooks which flow into it from their valleys.
For the first seven miles it bears east-south-
east,* and is about two miles wide, with precipitous, densely-wooded cliffs, rising to a height
of 500 or 600 feet directly from the water.
It then trends round to east-north-east, dividing into two arms, each about one mile wide.
These two arms enclose an island about twenty
miles long, and four or five miles wide in its
middle portion, which is very steep and equally
lofty with the surrounding country at its south-
west end, but becomes much lower at its north-

* The bearings in this Report are all " true," an allow-
ance of 25° having been made for the variation of the compass.

eastern extremity. From this island the pond runs in a north-eastern, and eventually in a north-north-eastern direction, and becomes wider as it proceeds, till it attains a breadth of about five miles. The surrounding country, especially that on the north-western side, gradually sinks towards the north into a very level and densely wooded tract, which extends towards the north-east as far as the eye can reach, no hills being seen in that direction except three low and distant peaks, which my Indian guide assured me were within six miles of some part of White Bay. At the north-eastern corner of the Grand Pond a considerable brook comes in, fifty or sixty yards wide, and several feet deep at its mouth, but three or four miles up it becomes too shallow for anything but a bark canoe. According to the information of the Indian, it proceeds from a chain of four ponds, each from four to six miles in length, the last of which is not more than fifteen miles from Hall's Bay, and by carrying their canoes about half a mile from which, the Indians meet another brook, down which they can float to Hall's Bay *.

* The mouth of this brook is marked in Bullock's chart "Indian Brook."

About three miles to the west of the mouth of the river in the Grand Pond, called "the Main Brook," an equally large brook runs out of the pond towards the north-west. This stream, marked in the map " Junction Brook," forms a succession of rapids for six or eight miles, when it falls into the Humber River. The Humber, from the account of the Indians, flows from two large ponds on the eastern flank of the "Long Range," and about in the latitude, or, as they expressed it, at the back of Cow Head. It is encumbered with several rapids, one of which occurs just above the mouth of Junction Brook, and for a quarter of a mile is of a very difficult and dangerous character. From this point the Humber runs to the south-west with an average width of sixty yards, and, though in places shallow, with a general depth of three or four feet. In about five or six miles it enters a pond called Deer Pond. This pond is fifteen miles long and three or four broad, lying in a north-east and south-west direction, and having its south-western extremity embosomed among the hills of the Long Range. Through these hills the river, escaping from the pond, finds its way by a narrow and precipitous valley into the Humber Sound. About half a mile from the pond is a dangerous rapid a quarter of a mile long: below this the valley expands to a width of about two miles, but within three miles of the mouth of the river the hills meet again, producing a rapid three-quarters of a mile long, but not of a difficult character. The scenery is here very picturesque; cliffs of white limestone rising in precipices of three or four hundred feet on the brink of the river, while thick woods of a finer character than usual conceal their bases, mantle over their sides, and frequently crown their heights. Through this narrow opening the drainage of a very considerable portion of the country is effected, extending from the latitude of St. George's Bay to that of Cow Head, and from the watershed of the long range of hills over a width of thirty or forty miles at least, into the interior.

The Indians informed me that a walk of about thirty miles due east (by compass) from the middle of the Grand Pond brings them to the southern end of the Red Indian Pond, and that a similar walk from that point to the south-east (by compass) takes them to the mid-

dle of another large pond, which must, I think,
be one of those called, by Mr. Cormack, Bathurst
and Jameson's Lakes : * they said, however, that
both the Red Indian and the other pond emptied themselves into the Bay of Exploits.

I did not succeed in getting more than
twenty miles up the River Exploits: the following sketch, therefore, of that system of
drainage is derived chiefly from the information of Mr. Peyton of Toulinguet.

The Red Indian Lake is about thirty miles
long and in the middle six or eight miles wide.
Its general direction is north-east and south-
west, bending, however, slightly at each end
towards the west, like the Grand Pond. At its
south-western extremity a considerable brook
flows in, which, according to Mr. Cormack's
map, proceeds from a large pond called George
the Fourth's Lake. Another considerable brook
flows in at its northern extremity, which comes

* The Indians assured me that the general direction of
these three ponds was the same, or nearly north-east and
south-west. In Mr. Cormack's map, Jameson's Lake lies
east and west, but he may have neglected in this instance to
allow for the variation of the compass. Bathurst Lake lies
about north-east and south-west, but it empties itself to the
south instead of the north.

from the neighbourhood of Hall's Bay. The pond contains two islands near its south-west shores, one nearly at its southern extremity, and several smaller ones in a bight on the south-eastern side. The land along its western and southern shores is bold and lofty, and the cliffs of its eastern shore south of the River Exploits are likewise high and precipitous. About ten miles from the northern end of the pond, on its eastern side, the River Exploits runs out. Its course is at first north-east, gradually bending as it proceeds more towards due east. The river is shallow and rapid, and for the first twenty miles receives but few tributaries. The first accession of any importance is the Badger Bay Pond Brook. This runs from two large ponds called Twelve-mile Pond and Badger Bay Pond, near the north coast, * passes through a succession of smaller ones round the northern extremity of Hodges Hill, and falls into the River Exploits a few miles west of that ridge. Two or three miles east of this a considerable

* With a flat or canoe on these ponds, this would probably be an easier route to reach the Red Indian Pond, than by ascending the River Exploits from its mouth.

brook comes in from the the south, called
Sheernock or Shannoc * brook, flowing from a
pond called by the furriers Sandy Pond. Eight
or ten miles east of the ridge of the Shute-
brook Hills, the river takes a sudden turn to
the south, and leaps down some falls fifty or
sixty feet over hard and craggy rocks. Below
these it whirls with a succession of pools and
rapids through a precipitous ravine for upwards of a mile, having cut back a channel to
that extent. Nearly half way down the ravine a brook comes in from the south, which
not having cut back a channel so fast as the
main river, shoots out its water from a height
of several feet like a spout or shute. It is
hence called " Shute-brook," and the hills
among which it rises the "Shute-brook Hills."
In ascending or descending the river the boats
or canoes have to be carried for a mile and a
half through thick woods to avoid the falls.
At the bottom of the ravine the river resumes
its easterly course, running nearly east-north-
east, thence to the sea, a distance of twenty

* Shannoc was the name by which the Red Indian desig-
nated the Mic-Macs.

miles. About eight miles from the ravine a
large brook comes in from the south, called the
Great Rattling Brook. This brook has a very
straight course for nearly thirty miles, running
in a northerly direction, and receiving several
smaller brooks and tributaries from the east
and west. Six miles below the Great Rattling
Brook, some smaller falls, called the Bishop's
Falls,* oblige the boats to be again carried
for 200 yards. Between the Bishop's Falls
and the Great Falls the river is generally from
100 to 200 yards wide, but very shallow,
and for the most part a succession of rapids,
full of large boulders and ledges of rock. A
short distance below the Bishop's Falls a ledge
crosses the river, where a rapid is alternately
formed and obliterated by the rise and fall of
the tide, which reaches to this point. Below
this the river is navigable for any kind of row-
boat, but above it nothing drawing more than
four inches of water can be got up in the
summer, and that only by tow-lines, and with
infinite labour and difficulty. The whole
length of the river from the lake to the sea is

* From the present Bishop of Nova Scotia having visited
them.

reckoned at sixty miles, being probably forty in a straight line.

By a salmon-fisher in Gander Bay I was told that Gander Bay Pond is about thirty miles long, but never more than two miles wide. It is crescent-shaped, the eastern end running to within about eight miles of the mouth of Travers's Brook in Freshwater Bay, while the west end trends greatly to the south. At its southern extremity two brooks come in, the most westerly of which is said, by the Indians, to come from near the Bay of Despair. From about the middle of the pond a brook runs out into Gander Bay, passing through three other small ponds, and having a very winding course, reckoned by the furriers at forty miles. From the numerous rapids and small falls, it requires three days to get a small punt up to the pond in the best season, one day being amply sufficient for returning.

There are two considerable brooks emptying themselves into Bonavista Bay, one at the head of Freshwater Bay, and the other at the head of the middle arm of Bloody Bay. The first is called Gambo Brook: it flows by a wide

but shallow and rapid channel two miles in
length, out of a pond which is one mile wide
and about nine miles long; at the end of this
another shallow and rapid stream of only half
a mile flows out of a second pond similar in
size and shape to the first. These ponds lay
in a pretty straight line bearing west-south-
west. Two small brooks flow into the upper
one at its farther end, neither of them navigable for anything but a bark canoe. From an
elevation two miles beyond the end of the
upper pond, nothing was seen but the undulating ridges and woods before mentioned as
forming the country to the west of this part of
Bonavista Bay.

Bloody Bay Brook discharges a very considerable quantity of water, and is upon the
whole the most important in Bonavista Bay.
Its navigation, however, is impeded at the distance of half a mile from its mouth by a very
rocky and dangerous rapid of more than half a
mile in length. Above this is " steady water,"
for six miles, navigable for a punt. I under-
stood from a salmon-fisher, the only person in-
habiting the neighbourhood, that a succession
of " steadies," with occasional rapids, may be

met with for twelve miles farther. There is then a fall, the water shooting clear over a precipice estimated at eighty or one hundred feet high. Immediately above this is a long and narrow pond called " Terra Nova Pond." This is reckoned to be twenty miles long, lying, as does the valley of the river, in about a true south-west course.* About six miles above the mouth of the river a brook flows in from a large pond on the north, called Maccles Pond, the northern end of which is within three miles of the middle of Lower Gambo Pond.

According to Mr. Cormack's map, there is a large brook running out of a considerable lake, called Barrows Lake, into North Bay, Fortune Bay. Another, emptying itself into East Bay of Bay Despair, runs through a chain of ponds from the immediate neighbourhood of Jameson's Lake. South-west of Jameson's Lake is another considerable one, called Bathurst Lake, emptying itself through a chain of

* I regretted much that, never having heard of this river and lake before arriving at its mouth, I had no means of ascending it ; as, if a sufficient quantity of provisions could be conveyed to the head of Terra Nova Pond, it would be a very favourable point for making excursions in the interior of the country.

ponds into Little River Bay; and from its
immediate neighbourhood another brook and
chain of ponds runs out to White Bear Bay.
It is by this latter route that the Indians informed me they proceeded into the country
when they wished to cross from the south
shore to the Bay of Exploits.

Great Codroy River (north of Cape Ray) is
navigable for boats as far as the tide runs up,
which is about nine miles; above that it is a
mere mountain torrent, rarely more than knee-
deep, and full of rocks and boulders.

The latter part of this description is applica-
ble to all the brooks flowing into St. George's
Bay.

In the rough map which accompanies this
Report, all the features now mentioned, as well
as some others of minor importance, have been
delineated. As, however, their positions are
laid down only from bearings and estimated
distances, nothing like accuracy can be claimed
for them. I have inserted Mr. Cormack's route
from a small map given in the ' Edinburgh
Philosophical Journal' for 1824, vol. x., page
156. As he used nothing but a pocket compass, his positions are, of course, only approxi-

mately accurate :* I have therefore, in one or two instances, shifted them a little, to make them accord better with my own.

* Under the circumstances of his journey, the very attempt to delineate the country reflects great credit on his industry, energy, and power of observation.

———————

SKETCH OF THE GEOLOGY

OF NEWFOUNDLAND.

Series of Formations.

THE aqueous or stratified rocks of Newfoundland consists of the following formations, which are arranged in descending order :—

Subdivisions.

1. COAL FORMATION. { *a.* Upper portion. { *b.* Lower or red portion.

1*a.* MAGNESIAN LIMESTONE.

2. UPPER SLATE FORMATION. { *a.* Belle Isle shale and gritstone. { *b.* Variegated slates.

3. LOWER SLATE FORMATION. { *a.* Signal Hill sandstones. { *b.* St. John's slate*.

4. GNEISS AND MICA SLATE FORMATION.

---

* In adopting these names I have been guided by the desire of only describing what I have seen, and especially of avoiding the use of terms which may have a theoretic import, until their propriety is proved. The general series of the stratified rocks of North America has first to be worked out on its own basis, before it can be advantageously compared with the European series, and the relative signification of the terms of the two satisfactorily settled. In the absence of all direct evidence of their place, then, in the general series, either from organic remains or the more un-equivocal proof of superposition, I have applied to the

The unstratified or igneous rocks consist of various kinds of Trap, Greenstone, Serpentine, Hypersthene, Porphyry, Sienite, and Granite. Erratic blocks, diluvial drift, or superficial accumulations, will be spoken of lastly.

1. THE COAL FORMATION.

The upper part of the coal formation consists principally of dark shales, clunch, or bind, with brown and yellow sandstones or gritstones in thin beds. It contains several small beds of coal, one of which has a thickness of at least three feet. This upper portion is similar to the coal-measures of England, but differs from them in containing some beds of red marl, or clunch, and in the sandstones frequently passing into conglomerates containing white quartz pebbles as large as an egg. The lower portion of the coal formation is characterized by beds of red sandstone, and red and green marls and gypsum. The two portions pass by perfectly insensible gradations into one another: yellow, brown, and

rocks of Newfoundland local names, rather than attempted to identify them with European formations by terms which have a chronological signification.

whitish flags and sandstones, dark blue clay, and an occasional bed of black shale occur throughout the formation, but are more abundant in the upper part, where alone they appear to contain coal. In the lower portion, the red colour gradually becomes more frequent till it predominates over the others. The red sandstones are generally soft and friable, and frequently contain quartz pebbles: they are also often blotched and streaked with green or white. Some of the lighter-coloured sandstones contained carbonate of lime, while in a bed of blue clay, crystals of selenite were observed; and in the red and green marls are large and important masses of gypsum.* This mineral occurs either as large veins of fibrous gypsum passing irregularly through the marls, or as thick beds interstratified with the marls. This latter variety is soft, powdery, and finely laminated, little black patches or very thin scales marking the lamination. Its formation seemed to be due to the disintegration of pre-

---

* This portion of the coal formation is so similar to the new red sandstone of England, that I was at first sight tempted to give it that name, till further investigation showed that it lay below the coal, instead of above it.

viously existing masses of gypsum and black shale, and the tranquil deposit of the debris in calm water. It is very abundant, occurring in several thick beds.

The total thickness of the coal formation must be considerable; neither its base nor its highest beds were seen, while the portion examined certainly had a thickness of 1000 or 1500 feet. It is denoted in the map and sections by a red colour, with streaks of black to mark the existence and position of beds of coal.

1 *a*. Magnesian-Limestone. No evidence was found to show the relation of this rock to the coal formation. The portion more particularly examined had a thickness of about fifty feet, in beds of from two to three feet each, frequently splitting into flags. It contained one bed of carbonate of lime, of a grey colour, two feet thick, with a band of brown chert. The magnesian limestone had generally a yellow colour, but rudely spheroidal concentric stripes of pink frequently occurred. These, in whichever direction the rock was split, produced markings similar to those seen in fortification agate, but on a much larger scale, being often two or three feet

across. No tendency to break or decompose
along the line of marking could be observed.
This rock had a much greater thickness on
the whole than the fifty feet mentioned above.

2. The Upper Slate Formation.

This formation was not observed anywhere
in the immediate neighbourhood of the coal
or magnesian limestone. The exact relations
between them, therefore, cannot be ascertained.
Every consideration of analogy, however, leads
us to believe these slates to be below the coal
formation in the series, though at what distance is not known. The upper slate formation consists of two subordinate groups, which
graduate or pass by insensible degrees into
one another.

a. The upper portion or group, called the
Belle Isle shale and gritstone, consists of dark
micaceous shale, with interstratified beds of a
very fine-grained grey gritstone. The shale
frequently splits into laminæ as thin as paper,
and when exposed to the air, rapidly decom-
poses into a very fine mould or dust. Sometimes, however, it is entirely composed of sil-

very mica, and is then more firm, and often corrugated like mica slate. On the firmer pieces occur singular markings in relief, which sometimes assume the shape of leaves, branches, or other organic bodies, but which are, I believe, entirely concretionary, and not organic. Beds of reddish stone, and of red marl or shale, of an inconsiderable thickness, occur sometimes among the upper parts of the shale. The interstratified beds of grey gritstone are generally about one foot thick, with smooth surfaces, and are much jointed by planes, which are frequently at right angles to the beds. In the upper part of this group the shale predominates, a few beds of stone occurring together here and there, with many feet of shale above and below them. As we descend to the lower part, however, the beds of stone increase in number and thickness, until the shale becomes subordinate to the gritstone, serving but as an occasional separation to some of its beds.

b. A little lower still, the shale disappears altogether, and we have then a mass of grey gritstone, frequently thick-bedded and generally fine-grained, but sometimes coarser, and

in one or two instances passing into a conglomerate of black and white quartz pebbles.
Sometimes, more especially when the beds
are fine-grained, the grey colour is variegated
with red, and the beds pass down into a mass
of slate of a brick-red colour, with a fine
cleavage. This cleavage is that of true slate,
crossing the beds at various angles, and being
perfect only in the finest beds, gradually dying
away as it approaches a coarser band. The
slate produced, however, is invariably brittle,
and unfit for economical purposes. From its
variety of colour, the group is called the
variegated slate. The colours of red and
greenish grey are capricious in their extent,
and sometimes calcareous beds of a brown
colour, with nodules of grey limestone, or
cream-coloured bands of slate rock slightly
calcareous, may be observed.

The total thickness of the upper slate formation must be very considerable, as each of
its two portions are many hundred feet thick
at least. As there does not occur, however,
any continued section in which the whole formation may be observed, it is difficult to
estimate its total thickness.

3. The Lower Slate Formation.

As no locality is known in which the lowest beds of the upper slate formation appear in contact with the upper beds of the lower slate formation, their precise relations cannot be determined. It is possible that they might pass into each other, or, at all events, that they are consecutive formations, in other words, that their antiquity is not greatly different. One or two clear instances, however, will be produced, in which beds of variegated slate rest uncomfortably on beds belonging to the lower part of the lower slate formation; proving some movements, at all events, to have taken place between the periods of their respective deposition. As, in this instance, it is not a question of practical importance, the lower slate formation may be considered as *next* in the series below the upper slate formation. The lower slate formation may likewise be considered as composed of two groups.

a. The Signal Hill sandstone consists of a mass of dark red sandstones and conglomerates. These are very hard, having a dull fracture, and are incapable of being easily worked into shape. The embedded pebbles are generally small, never being larger than a man's fist, and consist almost entirely of quartz. The beds are usually about three feet thick. The lower part of these red sandstones contains bands, or thick and irregular beds, of light grey gritstone, very fine-grained, and intensely hard, with a splintery and conchoidal fracture. It resembles the gritstone-beds of the Belle Isle shale in appearance, but occurs in much thicker beds, with fewer joints. This mass of red and grey sandstone, which, from its forming the hills at the entrance of St. John's Harbour, is called the Signal Hill sandstone, has a thickness of at least 800 feet. It passes down by a regular gradation into the slate rocks below, which, as the town of St. John stands upon them, are called,—

b. The St. John's slate. Beds of red, green, and grey stone, of a fine grain, alternate, near the junction of the sandstone, with the slate rocks, forming the transition beds between the two. These gradually get more slaty and perfectly cleavable as we descend. The cleavage of the slate is frequently parallel to the plane of stratification, more especially in its upper portions. In other places, however, the cleavage cuts the beds at various angles, and sometimes exhibits a beautiful "stripe" of blue, pink, and green. The thickness of the formation must be very great, certainly 2000 or 3000 feet, and probably much more. Neither this, however, nor the order of succession of its lower beds, can be sufficiently ascertained from the want of a good continuous line of section. There are beds of gritstone and large masses of conglomerate at various depths below the Signal Hill sandstone. Large thick-bedded masses of very hard, grey, fine-grained rock also occur without any cleavage whatever. In other places a well-developed cleavage produces excellent roofing slate. This formation is frequently traversed by veins of white quartz, and masses of porphyry are found associated with the slates, but whether of contemporaneous production, or as erupted and intrusive masses, is not often determinable.

4. Mica Slate and Gneiss.

The rocks of this formation in Newfoundland (supposing the term formation to be properly applied to them) do not differ from those of other parts of the globe. The mica slate, however, does not here appear to be separable from the gneiss, as they alternate with and pass into each other. Masses of quartz rock, chlorite slate, primary limestone, and the usual accompaniments of the formation occur also abundantly.

With the exception of some very indistinct vegetable impressions in the coal formation, I have never succeeded in discovering organic remains in any rock of Newfoundland. I have several times searched diligently, more especially in those parts that were at all calcareous. The limited extent of the exposed sections, however, the difficulty frequently of landing on the sea-cliffs, and the nature of the survey necessarily prohibiting long-continued examination of small spots, all combined to prevent that minute and accurate search which is probably necessary to find them if they exist.

The igneous rocks of Newfoundland do not, of course, differ from those found elsewhere: their several varieties, therefore, will be mentioned in describing the localities where they are found.

In entering on this description, we will begin with the province of Avalon.

Nearly the whole of the province of Avalon
is composed of the lower slate formation.
Along the eastern shore, from Cape St. Francis
to Cape Race, the beds of this formation have a
general easterly dip, varied, however, by numerous minor undulations. (See section, No. 2.)
By reason of this easterly dip, the headlands
which project farthest to the east are composed of the highest beds, and the land between Shoal Bay and Torbay, projecting some
miles beyond the general line of the coast,
consists of the upper part of the formation
only, namely, the red sandstones and conglomerates called the Signal Hill sandstone.
These beds are very well exhibited on the
south side of Torbay, and in the narrows of
St. John's Harbour. They form the ridge of
the Sugar Loaf, Signal Hill, and the South
Side Hill, and along this line they are inclined
to the east at an angle of 70°. On both sides of
the tongue of land forming Cape Spear,
however, namely, in Deadman Bay, and Petty
Harbour Bay, this easterly dip may be seen
gradually to decrease, and the beds, after be-

coming horizontal for a short distance, rise towards the east, into the cliffs forming Cape Spear, where they have a westerly inclination. This is a good example of a synclinal curve, the synclinal line running through the headland in about a north-north-east course. In going down any of the inlets and harbours along the eastern coast, the slate rocks in the cliffs exhibit a beautiful series of synclinal and anticlinal curves on a small scale, being continually undulated into regular arches, looking frequently like mason-work. On the south side of the inner cove of Torbay, and on the north side of Aquafort inlet, excellent examples of these may be seen. (They are similar, but on a much smaller scale, to those shown in section, No. 2.)

Either from the effect of these undulations, or from faults, the line of strike (or direction of the beds across the country) is sometimes, also, varied or undulated. The strike of the Signal Hill sandstone from Torbay to Shoal Bay is nearly north and south, but just below Shoal Bay it becomes north-north-west and south-south-east: its lower beds consequently soon come out to the sea-cliffs, which thence into the Bay of Bulls is composed of slate rock. South of the Bay of Bulls, however, the red sandstones come in again, dipping south-east and striking south-west, and Gull Island and Green Island are composed of them. By another flexure, and by the trending away of the coast, they are again thrown out, and their lowest beds are finally seen at the base of Cape Broyle, and in Ferryland Head, where they are nearly perpendicular. (See map.) The St. John's slate, then, forms the coast uninterruptedly to the southward, round into Trepassée Bay. On going into the country, from Renews to the hill called the Butterpots, slate rock is found the whole of the way to the flanks of the hill. It is then perceived to wrap round the base of the hill on its eastern and southern sides, and dipping everywhere from it at various angles, it apparently abuts against the porphyries and sienites which form the principal mass of the hill. The porphyry is a dark green rock, with a few disseminated crystals, and forms the exterior and summit of the hill: it apparently passes downwards or internally into sienite, as at the foot of one or two small precipices sienite of a reddish colour, and with rather large crystals, was found; the top of each cliff consisting of porphyry. (See section, No. 4.) No dykes were seen traversing the slate, but such may exist hidden beneath the thick moss and tangled woods on the slope of the hill. Near the summit of the hill curious patches of an apparent conglomerate of small angular pieces of porphyry were seen, very hard, and adhering firmly to the mass of the hill. It might be a conglomerate, or a mass traversed in every direction by small reticulated veins. From the summit of this hill, the range of broken and hilly ground mentioned before was seen to stretch to the northward towards Conception Bay, evidently composed for the most part of the same igneous rocks as the Butterpots at each end of it, while the slate rocks swept round to the south, producing generally a more level country, but forming also some lower hill to the west of the principal range. No other part of this range was traversed till it comes out on Conception Bay, where the Holyrood Butterpots is found to be of a precisely similar character to the Butterpots of Renews, namely, red sienite at the base, and for about two-thirds of the height of the hill, but capped by a mass of dark grey trap rock. No slate rock was observed on the western slope of the Holyrood Butterpots ; but, from the character of the country, the slate must approach to, or abut against, its eastern side. From the Holyrood Butterpots the St. John's slate formation runs down at a distance of about three miles from the shores of Conception Bay, till it approaches Topsail and Broad Cove, where it forms the sea cliffs. The bold height of Topsail Head is chiefly a mass of pure white quartz rock. At Portugal Cove the slate is found to be traversed by dykes or large veins of greenstone and trap rock, and masses of an impure serpentine occur, together with hard, grey, compact quartz rock. From the shapes of the hills, intrusive masses of igneous rock probably occur among the slate from Portugal Cove down to Cape St. Francis, where the slate is almost entirely concealed and supplanted by a close-grained siliceous porphyry. The slate rock in the neighbourhood of Portugal Cove and Topsail has generally the mineralogical character of grauwacke and grauwacke conglomerate.

These must be some of the lower beds. (See section, No. 1.) The higher beds at Middle Cove, Torbay, and in the neighbourhood of St. John's are fine slate, with a good cleavage, and might be used for economical purposes.

Round the head of Conception Bay igneous rocks predominate. On the east side of Holyrood is found a yellow, crystalline quartz rock, with circular nodules of a grey rock of inferior durability to the yellow quartz, in which, by decomposing more rapidly, it leaves a number of curious basin-shaped hollows, about two feet in diameter. The western side of Holyrood, down as far as Chapel Cove, consists of porphyry, and thence through Salmon Cove, Cat's Cove, and Collier's Bay, down to Turk's Head, and Bull Cove, the principal rocks are porphyry and sienite. Some slate rocks, to be mentioned presently, compose the extreme points of the headlands. Between Cat's Cove and Salmon Cove a red sienite occurs, which might be used as a building stone. The Cat's Cove hills are composed of porphyry passing into an amygdaloid of white crystals in a purple base. The western boundary of these rocks runs from the western

flank of the Cat's Cove hills straight to Bull
Cove, near Brigus, and the last trace of the
porphyry is seen in the cliff immediately
south of Brigus Harbour. To the west of this
boundary the St. John's slate sets in with a
high westerly dip, and runs thence uninterruptedly down the western side of Conception
Bay, as far as Bay Verde. Throughout this
space the general inclination of the beds is
towards the west-north-west: there are, however, many minor undulations, which may be
seen in the cliffs of the various inlets and
harbours which cut across the strike of the
beds. Very much of the slate rock immediately on the coast has a fine grain, and
would make excellent roofing slate. In Flam-
borough Head the slate rocks are perpendicular and appear disturbed; and in Bay
Verde they pass beneath a mass of red sandstones and conglomerates evidently belonging
to the Signal Hill sandstone. These sandstones form the whole of the headland thence
to Old Perlican, but the slate rock re-appears
in Baccalieu Island. From Old Perlican
to Heart's Desire the cliffs are composed of
the lower slate formation, but its details were

not examined. It is principally composed of slaty rocks, striking along the coast, and dipping generally to west-north-west. From Heart's Desire, towards the south, this formation trends away from the sea-coast, and leaves a strip of lower land between it and the sea, occupied by beds of the upper slate formation. The boundary of the two is obscured, partly by a great accumulation of gravel and boulders, and partly by marshes and woods.

Returning now to the south coast of Avalon, we find the St. John's slate on both sides of Trepassée Bay, and forming the whole country thence to St. Mary's and Placentia. Several large and regular flexures may be observed along the western side of Trepassée harbour and along the coast by Cape Pine and Cape Freels. The anticlinal and synclinal lines of these are strictly parallel to each other, and run along á true north-north-east course. Along the eastern shore of St. Mary's Bay, as far as the harbour, the dip of the slate rock is generally to the east, but in Mal Bay, after one or two more undulations, it becomes westerly at a high angle, and continues so thence to Salmoniez. The same beds are by the effect of these

undulations continually re-appearing, and a
description of one part serves for all.   Round
the head of St. Mary's Bay, about Colinet and
Rocky Rivers, some beds of hard thick conglomerate, generally grey or brown, with white
quartz pebbles, are found, as also here and
and there some thin beds of soft and rather
shaly rock.   There was no reason, however, to
suppose that these beds belonged to any other
than the lower slate formation; and the same
north-north-east and south-south-west anticlinal and synclinal lines were frequent, causing
the rocks to dip alternately to the west-north-
west and east-south-east, at various angles.   In
North Harbour the dip is westerly; and on
the banks of a small brook there, a jet-black
slate, with fine white laminations, was observed, dipping also to the west.   The range
of hills mentioned before as running from St.
Mary's to Trinity Bay consist of beds belonging to the lower slate formation, but rather
differing in character from those which are
generally seen elsewhere.   They are very
hard, rather coarse-grained, splitting into flags
rather than slates, brown or red outside, grey
inside, and dipping at a high angle to the

west. On North Harbour Lookout, and on the North-east Mountain, the sharp weather-worn edges of the beds bristle up from the bare ground along the strike of the rocks in a most singular manner, indicating both the strike of the rocks and the cleavage to the north to be 15° east, true bearings. West of this ridge, the country is obscured till we come to the head of the north-eastern arm of Placentia Harbour. Here a red sandstone and slate rock is found, dipping towards the east at a considerable angle, from beneath which other slate rocks, identical in appearance with the St John's slate, continue to rise towards the west of the shores of Placentia Bay. (See sections, 2 and 3.) At Point Verde, near Great Placentia, is a conglomerate in the slate formation, which contained, among many quartz pebbles, one about the size of a man's fist of a dark-red sienitic rock of a peculiar character, and identical in appearance with a rock forming a wide tract of country on the opposite side of Placentia Bay. Around Great Placentia there is much porphyry, principally dark-greenish-grey, with white crystals. The Castle Hill, on the north side of the harbour, and a considerable part of the cliff along the north-eastern arm, is all porphyry. On the south side the junction of the porphyry with the slate-rock may be seen in Dixon's Hill, which is a mass of porphyry with patches of slate-rock abutting against it. In approaching the por-phyry the slate loses its cleavage, becomes tough and rather crystalline, and at last passes into the porphyry by such insensible grada-tions, that it was only by carefully observ-ing the faintly-coloured stripe of the slate that the place of junction could be discovered. The porphyry stretches into the country to-wards the south; and, from the shape of the hills called Sawyer's Hills, a few miles south of Placentia, it is probable that they are chiefly composed of porphyry or other igneous rocks. With the exception of these, the whole peninsula between St. Mary's and Placentia Bays is composed of the lower slate forma-tion, which formation runs likewise through Little Placentia, as far as Long Harbour at least, and forms a great part of the isthmus that connects Avalon with the main land. In crossing from Long Harbour to Chapel Arm, in Trinity Bay, we find, on the banks of a brook about three miles from the latter place, a bright red-slate, evidently belonging to the upper slate formation, dipping towards the north, and passing upwards in that direction into dark shale. The shores of Trinity Bay, from Heart's Desire to Dildo Harbour, and thence through Chapel Arm to Tickle Harbour, are composed of the upper slate formation. Between Heart's Desire and Dildo Harbour the variegated slate-rocks, exhibiting a frequent alternation between greenish-grey and bright red colours, are traversed by several anticlinal lines, running in a direction, as nearly as possible north-north-east and south-south-west. In consequence of this undulation of the beds, patches of shale and gritstone are brought in here and there in the hollows of the slate-rocks, and the gradation from one into the other abundantly exhibited. One interesting locality where the shale and gritstone is shown is between Long Point and Witless Bay. In approaching it from the north, the red slate is observed dipping about south-east, and passing under some grey fine-grained gritstone or slate-rock, destitute of cleavage. As we successively come

upon higher beds, we find the gritstones gradually become separated by thin partings of
shale. These thin partings, as we proceed,
increase in thickness, and the beds of gritstone diminish, till we come to a mass of
shale, fifty feet thick, without any stone-beds
whatever. This shale is nearly black and
rather hard, but splits into fine laminæ, and
the cliffs are coated outside with great streaks
of brown and yellow, so commonly seen in
shaly cliffs. The shale lies in a beautifully
symmetrical basin or trough, rising from the
centre at an angle of about 45° on either hand.
Proceeding from the centre towards the south,
we find the same beds successively rising
into the cliff which were seen dipping below
it towards the north, until we come again to
the bright red slate-rocks. (See section, No.
5.) The effect of cleavage is very peculiar
here. In the lower beds of bright-red slate
the cleavage is well developed, the rock splitting across the beds into fine slate, which is,
however, very brittle. In the grey beds between this and the shale the cleavage is
scarcely perceptible, though they are not of
a sensibly coarser grain than the red rocks.

In the shale, however, the cleavage is again apparent. These beds are as perfectly fissile as any shale along the planes of lamination, and the laminæ separate readily whenever a portion is detached from the bed; but the mass of the shale is likewise traversed by a fine cleavage, preserving a constant angle of nearly 90° to the horizon, and having the same strike as the beds. The shale is thus minced, at it were, into small scales, or little narrow chips, being cut *thin* by the lamination, *narrow* by the cleavage, and thus made too fragile to retain any *length* in the direction of the strike of the beds. The lower surfaces of the gritstone-beds, alternating with the shale, are likewise traversed by the cleavage for an inch or two upwards, as they break or decompose into sharp jagged edges. About half a mile south of this spot, in a small cove opposite Red Rock, among the red slates, a band of red and brown calcareous rock was observed. It was traversed in every direction by small strings of carbonate of lime, and contained concretionary balls of grey crystalline limestone. Beneath was a pinkish-yellow con-

cretionary rock, with veins of carbonate of
lime and small balls of ironstone. The
thickness of these beds was about twenty
feet, and they are capable of being burnt
for lime. Some of the strings of carbonate
of lime looked at first so like fragments
of shells, that I searched diligently for
organic remains, but without success. Towards the head of the bay the variegated
slate formation sweeps into the interior for
four or five miles, forming a tract of land
more capable of cultivation than the generality
of the country. On entering Chapel Arm we
come immediately on some igneous rock.
This is for the most part a rather largely
crystalline greenstone; its texture, however,
sometimes varies into a nearly compact basalt.
It is frequently marked with circular bands
in relief of some inches in diameter: these
are sections of spheroidal concretions, which
are not, however, sufficiently developed to be
detached from the mass, and the nuclei of
which are of the same character as the rest
of the rock. On the west side of Chapel
Arm, near the Point, the variegated slate-rock
abuts against the greenstone without under-

going any apparent alteration, except that its colours become fainter, and that the red beds lose that hue entirely as they approach the greenstone. A little farther up black shale is seen, and beyond that greenstone again, and the two continue to alternate in the cliffs to the head of the Arm. The shale is hard and brittle, and rings with a metallic sound ; and the greenstone is apparently in the form of dykes or spurs, from the hills called the Tolt and the Monument, immediately to the west of the inlet. On the east side of Chapel Arm patches of dark shale and grey gritstone rest upon, and are caught in amongst, the greenstone, and are of course greatly altered from their original characters. The shale is hard, brittle, rings with a metallic sound, and does not easily split into thin laminæ, but rather into small flags, half an inch in thickness, though the marks of a much finer lamination are plainly visible externally. The gritstone is dark outside, almost crystalline in texture, and in places jointed so as to assume an irregular columnar form. The greenstone does not come out upon the coast at any other part of the neighbourhood, but the adjoining hills are probably formed of it, and from its conical shape it is possible, also, that Spread Eagle Peak, about five miles distant, is composed of the same rock. The red slate continues through Long Cove and Collier's Bay, where it dips to the north beneath the shale that forms the long tongue of land called Tickle Harbour Point. On the western side of Tickle Harbour Point, near its extremity, a great bed of grey conglomerate is seen in the shale, forty or fifty feet thick. The pebbles, consisting of white quartz, are seldom larger than walnuts, and are compacted together by a grey cement which is slightly calcareous.

In the districts now mentioned no clear evidence is given of the relation between the upper and lower slate formations. That they are not the same thing, however, is proved by the fact of the variegated slates passing upwards by regular gradation into black shale and grey gritstone, while the slate in the neighbourhood of St. John's passes upwards into a very thick mass of dark-red sandstones and conglomerates. It is highly probable also that the two formations are in Trinity Bay unconformable to each other, as nowhere in the neighbourhood of their junction are the upper beds of the one or the lower beds of the other to be seen, and therefore they cannot graduate one into another. If we return to Conception Bay, we shall find patches of a bright-red slate resting upon and apparently abutting against the lower slate at the points of Bay Roberts, Port-de-Grave, and Brigus Harbours. These patches of red slate dip towards the east, while the other slates have a westerly inclination; and in a cove just south of Brigus Harbour, called Sculpin Island Cove, the beds of variegated slate are clearly seen to overlap and cover the edges of the lower slate in a perfectly unconformable position. (See section, No. 6.) The lower slate dips to the north-west at an angle of 45°: it is rather thin bedded, of a dull-green and reddish colour. The upper slate is bright-red, thick-bedded, with bands of a cream-coloured concretionary rock and calcareous nodules, and dips at an angle of 30° to the north-east. The cleavage is well marked in the upper beds, which it traverses nearly at right angles, while in the lower it is faint and imperfect and appears to coincide with the strati-

N 3

fication. On the south side of the cove, part
of a dyke or other mass of porphyry is seen
cutting through and contorting the older slate,
but it is not observable on the north side of
the cove, which is occupied by the red slate.
These variegated slates, of which the bright-
red colour is most conspicuous, form the extreme points of the headlands of Collier's Bay,
Bacon Cove, Salmon Cove, and Holyrood,
resting sometimes against the porphyritic rocks,
and sometimes separated from them by a mass
of schistose and apparently altered rocks. Near
the western point, at the entrance of Holyrood, the red slates dip gently to the north,
but at the distance of 300 or 400 yards are
suddenly contorted, and turned up by a mass
of grey crystalline trap-rock or greenstone.
Near the junction of the two the colours of
the slate are much duller than usual, it becomes more brittle and siliceous, and loses its
previously well-defined cleavage. The continuation of this section to the south is obscured by loose sand and gravel, but in Chapel
Cove are some thin beds of limestone dipping
slightly to the north. This limestone is of a
grey colour, is very compact and siliceous, and of

a poor quality. It is traversed by small tubular
concretions of calcareous spar, which look as if
formed in the hollows where a vegetable stalk
or other organic body had decomposed. It is
not more than ten feet thick, and has both
above and below it some grey schistose beds
which are slightly calcareous. To which formation it belongs there is no direct evidence
to inform us. There is a strip of low land
running down from a little north of Holyrood
to Topsail, and having a width of from two to
three miles, left previously undescribed: this,
together with Bell Isle, Little Bell Isle, and
Kelly's Island, is composed of the dark shale
and gritstone. The tract on the main land is
covered with diluvial detritus, having a portion
only of the shale exposed here and there. The
perpendicular cliffs of the islands, however, expose every bed to the view. The general dip
of the beds of which the islands are composed
is to the north-west, so that the lowest beds are
seen on the south-east side. On the south-
east side of Kelly's Island a mass of gritstone
in many beds, having a total thickness of thirty
or forty feet, rises into the middle of the cliff;
and as the soft beds of shale on which it rests

have decomposed or been washed away, it has continually fallen down, so as to form a great mass of disjointed fragments at the foot of the cliff. This heap of fragments, being protected from the action of the breakers by a considerable pebble beach which stretches out around it, remains as a great natural stone-yard, from which much stone has already been carried away for the erection of the Catholic cathedral and other buildings in St. John's. In Little Bell Isle, as well as in Bell Isle itself, several bands of similar stone exist, but none of such thickness, nor in so favourable a situation for working, as in Kelly's Island. In the upper beds of Bell Isle, those namely on the north-west side, there is but little stone, although one bed of bright-red sandstone about eight feet thick was observed.

Several faults of greater or less magnitude are discernible in the cliffs of these islands.

Having given a sketch of the structure of Avalon, we will now proceed to trace the same formations farther to the west from the entrance of Fortune Bay, through Placentia, Trinity, and Bonavista Bays, to the Bay of Exploits. The French islands of Miguelon and

Langley are composed partly of variegated slate rocks and partly of reddish sandstones and slaty rocks belonging to the inferior formation. Langley Island is chiefly formed of red and purple slaty gritstones; an anticlinal line runs through it in about a north-east and south-west direction from Cape Percée. Immediately at the cape some bright-red slate rocks abut against the gritstones seemingly in an unconformable position, but on the western side of the island, near the neck of land connecting it with Miguelon, bright-red and grey slates dip to the west from the centre of the island at the same angle as the gritstones inside them, and pass underneath some shale; the whole lying in an apparently conformable position. St. Peter's and the mainland opposite, from Point Mary to Cape Chapeau Rouge, is composed of a dull-red igneous rock. This is sometimes a compact red feldspar, with here and there a whitish crystal, sometimes a regular red and white porphyry, and sometimes it passes into a red sienite of feldspar and quartz. It forms on the mainland a low country destitute of trees, and rising into low barren hills in the interior. At Cape Chapeau Rouge a

dark slate rock sets in, and runs along the
coast, through St. Lawrence and Burin, for
some distance along the west side of Placentia
Bay. It is very much broken and contorted,
too shivery to form roofing-slate, and dips in
various directions. In the small inlet of Mortier Bay there is a great and most perplexing
variety of rocks : the dark-green schistose beds
above-mentioned continue for about two miles
into the bay, but are suddenly replaced by
quartz rock in a large amorphous mass on the
south side of the bay, while on the north a
serpentine with bands of quartz comes in, and
over these lie patches of black shale, with thin
beds of grey gritstone, precisely like the Bell
Isle shale formation, but much twisted and
contorted : these latter rocks run for some distance on the north side of the bay, into the
large cove called Spanish Room. On the south
side of the bay, the quartz rock, after forming
a lofty cliff for about half a mile, suddenly
ends, and regular beds of variegated slate are
found abutting against it and dipping from it
in a westerly direction. The bay here trends to
the south-west, and these rocks apparently continue along its southern shore ; on the opposite

side of the bay a peninsula juts out, forming
the south side of Spanish Room ; it is nearly
a mile in length, and is composed of the following rocks. (See section, No. 7.) The
point of the peninsula is occupied by a rock
which whether to call it a sandstone or a gneiss
is a matter of doubt.* It has evidently been
formed of the detritus of red sienite, a round
pebble of which rock I found enclosed in it :
it is tough, but not very hard, it is regularly
bedded, dips to the north-west at an angle of
70°, and is divided into square blocks by joints
that follow the dip and strike of the beds.   It
would make a very fair building-stone, if care
were taken to place it with its planes of lamination in a horizontal position.   The thickness
exposed of this rock is about 200 feet.   To
the low cliffs composed of this succeeds a
small bank of sand and rubbish, immediately
beyond which is another cliff about forty feet
in height, composed of beds of red and green
marls, containing a band of red sandstone and
conglomerate, dipping at a very slight angle

---

* It resembles gneiss in its crystalline components, but
differs from it in these components being more loosely connected, the particles being more easily separated by a blow.

to the south-west, and exposing a thickness of
about 150 feet. In the lowest beds of marl
are bands of white marl indurated and very
calcareous, and one or two beds of very hard
concretionary limestone, mottled with red and
white. The cliff again ends, and a low bank
of sand and boulders extends for about 200
yards, when suddenly some black and brown
shale is found resting on two beds of light-
brown or whitish limestone, siliceous, and containing small tubular concretions and strings
of spar, and agreeing in every respect with the
thin beds of limestone in Chapel Cove, Holyrood, at the head of Conception Bay. The
two beds of limestone are separated by a thin
parting of shale ; they are each about five feet
thick ; and the whole mass of shale and limestone dips at an angle of 75° to the south-south-
east. The beds of limestone form a ridge running across the beach and keeping the same
dip and strike for some distance into the water.
Unfortunately the section here is again interrupted by a hollow filled with sand and
boulders, immediately beyond which is a cliff
of red sandstone and conglomerate, dipping in
the same direction with the red marls and

sandstones before mentioned, and exposing a thickness of about forty feet. This last mass of conglomerate is rather soft, full of large quartz pebbles, imbedded in fine red sand, and marked by regular lines of stratification. The remainder of the peninsula is a low beach running up to the main land, the cliffs of which are there composed of the same serpentine rock, associated with quartz, which was mentioned before. These beds of red sandstone and marl are certainly very like those belonging to the lower part of the coal formation. Their extent, however, is so limited that it is impossible to say to what they belong, and I never saw any other beds like them in the eastern portion of Newfoundland. I did not visit any other portion of the west side of Placentia Bay, but from its rugged appearance throughout, and from the structure of the neighbouring islands, I should judge it to be composed chiefly of igneous rocks.

The island of Audierne consists for the most part of a mass of dark-purple porphyry and quartz rock, against which a patch of variegated slate abuts at one part of the harbour.

The Isle of Valen consists of a red and grey

slate rock, with a grauwacke conglomerate of small angular quartz pebbles.

The island of Merasheen consists of a great ridge of porphyry running down the centre of the island, with patches of slate rock on each side of it. In Merasheen Harbour alternate bands of slate and gritstone are seen. The gritstone, where it approaches the central porphyry, becomes a white and rather compact quartz rock. The slates are black with a white stripe, and in places have a fine cleavage, and would probably make good roofing-slate. They dip to the north-west at a high angle, and the upper part contains some beds of dark-red sandstone and conglomerate. The Ragged Islands, on the west side of Merasheen, consist of porphyry and granite or sienite. The granite is confined to the low islands in the centre, and it sends large and frequent veins into the porphyry. Barren Island I did not visit, but, from a block I procured from it, it appears to be composed of steaschist. The northern part of Red Island is composed of granite or sienite. It is of a red colour, coarse-grained, principally quartz and feldspar, but with a few flakes of black talc or hornblende here and there. Near

the harbour are several bands of a grey rock
traversing the granite in nearly a north and
south direction.   This grey rock is finely crystalline, principally quartz, with a little hornblende.   The grey rock and the granite mutually intersect each other by veins in various directions.   I saw some pieces of a dark-
brown flag-stone from the south side of the
island, and heard of some red slate. Similar
red granite and grey quartzose rock to the
above compose the Ram Islands. The remainder of Placentia Bay I was not able to visit,
but from its appearance similar rocks to those
already mentioned run throughout it.   On the
north side of the neck of land dividing Placentia from Trinity Bay, the older slate rocks
are seen dipping to the north-west from Tickle
Harbour to Bay of Bull's Arm. A mile or two
west of Tickle Harbour is a mass of serpentine
with some obscure steatitic or feldspathic rocks,
apparently altered, and in one place a yellow
quartz rock containing crystals of feldspar.
Over these occur slate rocks of various character passing upwards in Bay of Bull's Arm
into purple and red gritstones, sandstones, and
conglomerates, which no doubt represent the

Signal Hill sandstones. These likewise dip to the north-west, and at the head of Bay of Bull's Arm they are overlaid by the variegated slate rocks. The cliffs hereabouts, however, are so low, and the junction so obscure, that it is impossible to say whether the two are conformable or not. From the Bay of Bull's Arm along the west shore of Trinity Bay as far as Buonaventure Head, the same rocks occur in the same position, the red sandstones dipping to the west under the variegated slates. A narrow band of these latter rocks runs from the head of Bay of Bull's Arm, through the centre of Random Island, into the country west of Pope's Harbour, and to the west of this band the red sandstones again rise to the west, and in Random south-west arm expose a long series of slate rocks, which rise to the west from beneath them, and are apparently similar to those near Tickle Harbour. The lowest of these slate rocks at the head of the arm is a smooth black shivery rock, very brittle, without any cleavage other than that of the lamination, and containing much hornblende and iron. This rock is likewise seen on the south side of Random Sound. In Random Island this recurrence of the red sandstones and inferior slate rocks is not apparent, the variegated slates being succeeded towards the west by the Bell Isle shale and gritstone, which form all the north-western corner of the island and a considerable tract on the mainland opposite it. A chain of hills with a sharply peaked and serrated outline runs through the country a few miles west of Bay of Bull's Arm and the inlets about Random Island. These hills send down a spur to the coast, opposite the western extremity of Random Island, which consists of red sienite: patches of schistose rock were at various places seen resting on the sienite, and at one point its junction with the shale was exposed. The sienite was found here partly to overlie the shale and gritstone in a slanting position, the beds dipping towards the sienite and abutting abruptly against it. (See section, No. 8.) No great alteration was observed, except that the shale might be a little firmer, and the gritstone more than usually hard and of a semi-crystalline texture. The detached islands about the mouth of Smith's Sound are composed of a red and grey fine-grained gritstone. In Anthony's Island the grey variety is full of large cubical crystals

of iron pyrites.   Just east of British Harbour
(called also Shut-in Harbour) a large trap-dyke
comes out on the coast, cutting through the red
and purple gritstones without producing in them
any sensible alteration or disturbance.   This
dyke is about 200 yards wide : near its sides
the rock is vesicular, and nearly black, the ca-
vities being here and there filled with white crystals.   Farther towards the centre of the dyke
it becomes compact, of a dark-grey colour, and
the central portion of the dyke is rudely columnar.   The part in which the columnar
structure is best developed is about twenty
yards wide, forming a nearly perpendicular
band slightly curved.   The columns are horizontal, very short, irregular in the number of
their sides ; and the outside ones are slightly
bent, those on the east downwards, those on
the west upwards.   (See Section, No. 9.) Near
the principal dyke two or three smaller ones
were observed cutting through the gritstone
without disturbing it.

From Pope's Harbour to Trinity Harbour
the lower slate formation, containing red and
purple gritstones and conglomerates of small
pebbles and greenish slate rocks, forms   the

whole country. One anticlinal line passes
through New Buonaventure and runs into the
country in a north-north-east direction: west of
this the rocks dip westerly, to the east of it
they incline to east-south-east at various angles. These same rocks continue from Trinity to Catalina and Bonavista, having in
the former place an easterly, in the latter a
westerly, inclination. Catalina is locally cele-
brated for an abundance of iron pyrites, which
is found there in a grey slaty rock of very fine
grain and frequently destitute of cleavage.
The iron pyrites occurs in large cubical crystals : it is called, in Newfoundland, Catalina
stone, and has often been mistaken for copper
or gold. In Bonavista the slate rock is ordi-
narily destitute of regular cleavage, but much
divided by lines in all directions. About Keels
a quantity of dark-red gritstone forms high and
barren ground, and is probably the continuation of that forming part of Random Island.
It represents the Signal Hill sandstones. It
dips rapidly in some places to the east, in
others to the west, and is traversed near the
west head by a mass of white rock, apparently
quartz. The Long Islands and their neigh-

bourhood are composed of greenish-grey slate rocks, which dip to the north-west. The deep inlet of Clode Sound, though it crosses the strike of the rocks, does not expose a very satisfactory section. Near its head, about Platter Cove, a dull-red feldspathic rock, similar to that found at St. Peter's and Lameline, runs along the east side of the south-west arm towards the south, and for some distance into the country towards the north. West of this the banks of the brooks exposed here and there masses of hard dark-brown slaty gritstone like that of the north-east mountain of Placentia, forming continual sharp ledges, over which the brooks fretted and foamed in their course towards the sea. In the river, which comes in at the head of the south-west arm, some beds of red and brown sandstone, in some places soft and shaly, were observed, with a slight dip to the west-north-west, and in another brook bright-red slaty rocks looked like beds of the variegated slate formation. East of the band of red igneous rock, which is about two miles wide, a red and yellow schistose rock, very rotten and crumbly, and consisting principally of chlorite slate, occupies some distance. Upon this in Long Cove and one other place rested some hard brown sandstones and conglomerates, which on the south side of the Sound seemed to run into the country for some distance. In Brown Cove and its neighbourhood the rock is a black slate similar to that at the head of Random south-west arm. Almost all the islands which fill Bonavista Bay, north of Clode Sound, are composed of slate rock, belonging probably to the lower slate formation. In Morris's island the slate was found to be black, with a few bands of grey gritstones; the slate had a fine cleavage, and would probably make a good roofing slate. It would be as useless as it would be tedious to enter into the detail of the position of the beds in all these localities: suffice it to say, the various dips and strikes of the beds and cleavage, when observed, were not found to vary from those seen in other parts of the country. In Bloody Bay and for five miles up the main brook of the middle arm of that bay, the prevailing rock was a grey slate with but little cleavage. The Lonil Hills, however, and probably Mount Stanford, are composed of a grey quartzose granitic rock,

rather fine grained and spotted with brown.
From Bloody Bay to Content Reach, grey
slate rock, similar to that just mentioned, is
alone found. At Man Point a dark red gritstone, however, is seen, and Man Point Ridge
is composed of a light greenish grey, fine-
grained, very hard gritstone, in thick beds.
It dips to the north-east. At the mouth of
Freshwater Bay a dark schistose rock occurs
in a perpendicular position, and strikes thence
through Hare Bay into Locker's Bay. The
shores of Freshwater Bay to the mouth of
Gambo Brook are all granite, which runs
thence down the west side of Locker's Bay to
Chalky Cliff.

In the country, however, round Gambo
ponds a dark slate rock is alone visible. This
slate rock differs sensibly from that found in
Avalon, or even in Bonavista Bay. It is
smooth, shining, splits along the planes of
lamination, has no cross cleavage, is sometimes micaceous, but consists principally of
chlorite slate. Its strike is nearly north and
south, with a general inclination to the west,
and its thickness must be very great. The
granite at Chalky Cliff is composed of large

crystals, sometimes reddish, but generally of a pale flesh colour, weathering almost white. The crystals of feldspar are very large and perfect. This granite is tolerably easy to break, and would make an excellent and very handsome building-stone. Many large loose blocks are now lying about near Chalky Cliff, and could readily be carried away in calm weather.

In Trinity Gut a sloping sheet of this granite was exposed, inclining at an angle of 35°, in size between 200 and 300 yards square, and, as far as I could see, without a joint or line of division of any kind whatever. This granite forms the Fair Islands, all the country round Indian Bay, the islands about Greenspond, and the mainland round Cape Freels as far north as Muddy Hole and Ragged Harbour. At Greenspond it is a very hard rock, of a grey colour, with large crystals of white feldspar. This variety would make an excellent building-stone, and have a very handsome appearance, as it gets whiter by exposure to the atmosphere. In Newell's Island, on the south side of Greenspond Harbour, some gneiss and mica slate is found, and the junction

o 2

of these rocks with the granite clearly exposed.
At some distance from the granite the rock is
thick-bedded mica slate, of a dark grey colour,
the mica being in large flakes. It contains
some long lenticular-shaped masses of a fine-
grained rock, nearly black, running along the
strike of the beds: this is almost entirely composed of very minute scales of mica. Approaching the granite, and about one hundred yards
from its boundary, small nodules and strings
of yellowish quartz rock are visible in the mica
slate. The mica slate then splits more easily
into thin laminæ, and is more varied in its
composition, some beds having more the character of coarse gneiss, interstratified with
others entirely composed of large flakes of
mica. On continuing to approach the granite,
the quartz veins and nodules increase, and
nodular patches and bands of a regular fine-
grained granite with but little mica appear.
These granitic portions are not veins proceeding from the neighbouring mass of granite,
but integral parts of the beds; a perfectly
laminated bed, gradually losing, first its fissile
character, and then its laminated appearance,
and passing in the direction of its strike into

a band of fine-grained flesh-coloured granite
several inches thick.   This band of granite,
after the course of a few feet, gradually thins
out again, and the bed regains its original
character of gneiss or mica slate.   This alternation and passing of one rock into another
increases in frequency, until, after walking
over the edges of many such beds, we find
ourselves imperceptibly led to a mass of red or
flesh-coloured granite perfectly crystalline, and
having no appearance of any lamination or
bedding whatever.   In the granite itself, however, for some distance from the junction, nodular masses of the black rock mentioned before as consisting of minute scales of mica were
observed.   This red granite does not contain
the large crystals of feldspar which are found
in the grey variety on the other side of the
little harbour, and is altogether more close-
grained than that rock.   Altogether the gradual passage or transition from the granite
into gneiss and mica slate was most remarkable.   The granitic portions contained in the
mica slate and gneiss struck me as just such as
would be produced supposing great heat to be
applied to a mass of rocks, some portions of

which were in a state to be more readily
affected and more thoroughly changed by it
than other and intermediate parts.

At one or two points along the shore north
of Cape Freels small patches of mica slate
and gneiss similar to these may be observed
resting on the granite. In a headland just east
of Cat Harbour the granite occurs in immense
sheets unbroken for twenty or thirty yards
by any division-line. The principal joints
were perpendicular and struck east 35° north,
quite parallel to each other and at regular distances of about twenty yards, no cross-joint
occurring sometimes for thirty or forty yards.
In one loose block, of very large size, a mass
of gneiss and mica slate with a fine lamination was enclosed on two sides by granite
with large crystals of feldspar. The junction
of the gneiss and the granite was well marked,
but the two were firmly united, and the large
disseminated crystals of feldspar were in some
instances half in the granite and half in the
gneiss and mica slates. One detached piece
of the latter also was enclosed in the granites.
I did not myself visit the western boundary
of this granitic district, but was informed that

it was situated near Ragged Harbour. In
Gander Bay a dark shivery slate was found,
and slate and slaty rocks stretch thence into
the Bay of Exploits, and form the chief part
of all the islands to the northward. These
slate rocks probably belong to that called the
lower slate formation, but it is possible that
in some instances chlorite or other slates of a
still inferior formation may occur.

The central portion of the island of Fogo,
about Hare Bay, is composed of a red sienite
or granite of rather a peculiar character, being
almost entirely composed of crystalline quartz.
North of this, about the harbour of Fogo, is
the common light grey fine-grained gritstone
or slate rock, generally hard and compact, but
having here and there an imperfect slaty cleavage : it forms the bold hills about the harbour,
rising 500 feet above the sea, and dips to the
north or north-east. The Change Islands are
blue or green slate, and from the Indian Islands
I saw a specimen which would form good roofing slate. The North point of New World
Island contains large beds of coarse conglomerate with little or no appearance of bedding,
but having the enclosed pebbles arranged in

regular lines. The remainder of the island is almost entirely slate rock. Toulinguet Islands are principally coarse slate rock, but on the east side of Toulinguet Harbour a white granite, rather fine-grained, shows itself. In the low and crumbling cliffs along the eastern side of the harbour, and for two or three miles to the south of it, a trap-dyke of a very remarkable character may be traced here and there. It is a dark-brown trap, rather soft and readily decomposing, and it is full of small crystals of mica. These are hexagonal, generally one-eighth to half an inch long, and split horizontally into thin plates. The general width of the dyke is not more than three feet, and at the place where it is best exhibited the rock on each side is a whitish quartz.

At the head of the Bay of Exploits, near Lower Sandy Point, the cliffs are composed of grey slate rock, full of small joints crossing each other at all angles, but having no well-defined cleavage. The river Exploits, in the first twenty miles from the mouth, exhibits the following rocks. At High Point, about three miles from the sea, is a dark-coloured porphyritic rock, occasionally amygdaloidal.

Just above, however, is a brown and reddish gritstone and slate rock, striking north-north-east, and dipping westerly for some distance. Several small undulations and contortions occur, the axes of which are parallel to the strike of the beds. At the Bishop's Falls is a pinkish-green slate rock with similar curves. These rocks, or others like them, reappear wherever there is a cliff, or the substrata are exposed to view, as far as the foot of the chief falls, mingled occasionally with beds of light-brown or salmon-coloured gritstone, or very fine-grained sandstone, intensely hard, and sometimes slaty. Similar rocks are seen just below the falls, with brown ferruginous sparry veins. Beds of bright-red slate rock are inter-stratified with them, precisely similar in external character to the variegated slates of Avalon. The strike of these rocks is north-east by north, with a westerly dip at a high angle, and the cleavage nearly always coincides with the stratification. The falls themselves occur in a very thick mass of dull-red sandstone, finely grained, very hard, thin bedded, and traversed by many small irregular joints with sharp jagged edges. This likewise dips to the north-west, at an angle of 45°. The falls on the river Exploits was the most westerly point I reached on the northern side of the island. From all I could learn, however, rocks not very different in character from those now described, composed the remainder of the country as far as the Red Indian Lake, and the coast as far as Hall's Bay.

The other portion of the island of Newfoundland, the geological structure of which was actually examined during the survey, is the south-west corner, from the Burgeo Islands to the Bay of Islands.

Though I did not land on any point between Cape La Hune and the Burgeo Islands, yet from the contour of the coast, and the description I was enabled to get, I can safely assert it to be composed chiefly, if not entirely, of granite. About the Burgeo Islands granite is the sole rock, with the exception of some patches of mica-slate and gneiss on one of the headlands. Three varieties of granite were observed: one white, rather fine-grained, with abundance of mica; another of a coarse grain, with less mica, and of a reddish colour ; and the third, which is by far the most abundant, a somewhat

coarse red granite with large embedded crystals
of flesh-coloured feldspar.   These rocks occupy
the whole coast, and a wide tract of the interior,
between Burgeo and La Poile Bay.   Both the
east and west points of La Poile are composed
of the porphyritic granite mentioned above, or
that which contains the large crystals of feldspar.   On the east side of the bay this granite is soon replaced by porphyritic greenstone, which runs up to Galley Boy Harbour.
On the western side of the bay, however, the
granite runs up as far as Tooth Head, where it
partly overlies and sends large veins into a
mass of dark-blue and purple schistose rock
with a green stripe.   The changes which take
place at the junction of these two rocks, in
their respective characters, are worth observ-
ing.   At about ten yards from this junction
the imbedded crystals of feldspar in the granite
become smaller, and soon cease to be conspicuous, the rock is then principally composed
of crystals of quartz and hornblende.   The
portion from which the veins arise soon loses
the hornblende, the quartz from a crystalline
state becomes compact, and the veins at a short
distance from the granite are entirely composed

of compact quartz rock on the one hand, while their gradation into granite on the other is well and clearly exhibited. The granite itself also becomes more and more largely granular and crystalline as we advance into its mass. (See section, No. 10.) The schistose rock at its junction with the granite is hard, brittle, and traversed by strings of quartz: as we recede from that rock, however, it passes into a compact flagstone, in thin beds of a fine grain, hard but tough, of a light-green colour, occasionally having a slaty cleavage, when it resembles the St. John's slate. Its general dip is about south, or toward the granite, at an angle of 80°. About one mile above Tooth Head, in a large cliff of regular flagstone, without slaty cleavage, two granite veins are seen four or five feet across, whitish, consisting of crystalline quartz, feldspar, and hornblende, and producing no apparent alteration in the neighbouring rocks. On the eastern side of the bay opposite this is a mass of dark siliceous schist, with brown ferruginous stains, which is succeeded towards the south by quartz rock and chloritic schist, continuing to the greenstone porphyry mentioned before. I

was informed that slaty rocks were traceable
for several miles into the country beyond the
head of La Poile Bay. Between La Poile and
La Moine the rocks are all granite, principally red, and some of it of a rather fine grain.
From La Moine to the Dead Islands and
thence to Port-aux-Basques and Cape Ray,
mica slate and gneiss compose the entire country. About the Dead Islands abundance of
veins exist in the gneiss, some of which are
thirty yards wide, and are composed of large
crystals of quartz and feldspar, containing nests
of mica or hornblende; thus constituting a
very largely crystalline granite. These veins
always run with the strike of the beds, and
their sides present no well-marked line of
division between the crystalline rock and the
schistose mica slate and gneiss, one passing
into the other by fine gradation. Some well-
marked distinct granitic veins, however, were
observed, which not only ran in the strike of
the beds, but crossed them and enclosed masses
of the mica slate. No large mass of granite
appeared in the neighbourhood of these veins,
but such might exist a little way in the interior. The mica slate and gneiss do not occupy

distinct tracts, but beds of each alternate with
the other, and some beds partake of the character of both. The strike of these rocks is
everywhere pretty uniform about the Dead
Islands and Port-aux-Basques, being about
east-north-east; the dip, however, is northerly
at the Dead Islands, and southerly at Port-
aux-Basques. At the latter place, beds of a
peculiar character were interstratified with the
gneiss and mica slate. They were not more
than a foot or two thick, but were fine-grained,
black, heavy, and crystalline, consisting almost
entirely of small crystals of hornblende. Gar-
nets occur sparingly scattered about the mica
slate, but I observed none of any magnitude.
These gneiss and mica slate rocks continue
from Port-aux-Basques round Cape Ray, for
some distance towards Little Codroy river,
where they terminate. The chain of hills
called the Long Range running into the country from Cape Ray seems to be throughout
composed of gneiss and mica slate. The hills
are found to consist of these rocks in the country east of St. George's Bay and about the west
end of the Grand Pond, and also on the bank
of the Humber River, in the Bay of Islands.

The south side of St. George's Bay, between this range and the sea, is occupied by the coal formation. The cliffs on the sea-shore, and a band of country, of a few miles in width, parallel to it, exhibit the lower beds of the formation, viz. the red sandstones and marls with gypsum. In the cliffs near Codroy Island is much red and green marl, with bands of white flagstone. The white flagstone and the greenish marl contain many veins of white fibrous gypsum, and interstratified with these and the red marls are some thick beds of white and grey gypsum of a singular character. These gypsum beds are not hard compact sulphate of lime, but are composed of white flakes of that substance, regularly laminated, and interspersed with small flakes and specks, or sometimes thin partings of a black substance, apparently bituminous shale. The whole mass is soft and powdery, thick-bedded, and in considerable abundance, and it might be carried away in boats with great facility. I was informed by some Indians of Great Codroy River, that they had seen a bed of coal two feet thick, and of considerable extent, some distance up the country.

Their accounts of the distance, however, varied from ten to thirty miles, and I could not induce any of them to guide me to the spot. I proceeded up the river about twelve miles from the sea, and some distance beyond the part navigable for boats, without seeing anything but beds of brown sandstone and conglomerate, interstratified with red marls and sandstones, gradually becoming more horizontal, and dipping towards the south-east. I believe, however, that a bed of coal had been seen by an Indian on the bank of a brook running into Codroy River, about thirty miles from its mouth, but the person who saw it was not in the neighbourhood at the time of my visit. About the middle of the south side of St. George's Bay, in the vicinity of Crabb's River, the lower part of the coal formation, consisting of alternations of red marl and sandstone, strikes along the coast, the beds dipping to the north west, at an angle sometimes of 45°. About three miles from the coast, however, an anticlinal line occurs, preserving the same strike as the beds, or about north-east and south-west, and causing those to the south of it to dip to the south-east. Thus the rocks which form the country along the coast, to the width of three miles, with a north-west dip, again occur, of the same or a greater width, according to the angle of their inclination, with a dip to the south-east, before we can expect to find any higher beds than those in the sea-cliffs; so that at least six miles of the country, formed of the lower beds, must be crossed directly from the coast, before we arrive at the higher beds in which the coal is situated. (See section, No. 11.) In ascending the brook next above Crabb's River, I found on the sea-coast beds of soft red standstone and red marl; and half a mile up the brook red and whitish sandstones, interstratified with beds of marl, chiefly red, but also occasionally whitish, green, or blue; beyond that were beds of marl, containing massive grey gypsum, similar to that at Codroy, and a bed of blue clay, containing crystals of selenite. Similar rocks, with now and then a bed of brown or yellow sandstone, occurred throughout the first two or three miles, all dipping north-west, at various angles of inclination. Beyond this point, the dip was in-

variably south or south-east, and for two or
three miles farther the character of the rocks
was precisely similar to those I had already
passed. As, however, the banks of the brook
were occasionally low, the section observed
was of course not perfectly continuous, and
beds which were hidden on one side of the
anticlinal line formed cliffs, and were thus
exhibited on the other side. Thus, as I continued to ascend the brook, I came on a cliff
of red marl fifty feet thick, with some thin
grey soft micaceous sandstone, beyond which
were some beds of grey, hardish rock, with
nodules of sub-crystalline limestone, the banks
of the river being likewise covered with a
crust, a foot thick, of tufa. Some distance
above this, the red sandstones become more
scarce, the colour being generally brown or
yellowish; grey clunch, too, with bituminous
laminæ, was frequent. In one bank of brown
sandstone, a nest of coal, with a sandstone
nucleus, was seen. Its shape was irregular,
being about two feet long, and it was probably
the remains of some vegetable squeezed out of
all semblance of its former shape. Over this
mass of sandstone there was again a good

thickness of grey clunch, and brown or yellow
sandstone and conglomerate, interstratified
with red and brown marl, all dipping gently
to the south-east. Over these were some thin
beds of red sandstone, with red marl; and a
little beyond, some hard, light, brown, or
greyish-yellow sandstone, with small quartz
pebbles. This rock formed ledges, stretching
across the river, producing a fall of two or
three feet. About 150 yards above this, on the
west bank of the brook, was some grey clunch
and shale, on which rested a bed of hard,
grey sandstone, eight feet thick, covered by
two or three feet of clunch and ironstone-
balls, and two feet of soft brown sandstone, with
ferruginous stains, on which reposed a bed of
coal, three feet thick. (See section, No. 12.)
The dip of these rocks was very slight towards
the south, in which direction the bank became
low, as it was also on the opposite side of the
river, which prevented my tracing the coal farther; neither was the bank above the coal high
enough to bring in any of the beds over it,
and thus give its total thickness; since it is
evident the portion here seen may be only
the lower part of a bed, instead of the whole.

The quality of the portion thus exposed was
good, being a bright caking coal. The distance from the sea-shore is about eight miles;
the only harbour, however, is that of St.
George, which is about twenty miles from this
spot. A few very rude and imperfect vegetable
impressions were all I could see in any of these
rocks. Many of the gritstones in this section
might probably turn out good freestones. In
the next brook to the east of the one I as-
cended there had been formerly a salt spring,
which, however, I was assured had lately become quite dry, although several of the little
rills which I tasted in the neighbourhood
were brackish. As regards the extent of
country occupied by this bed of coal, or others
which may lie above it, the data on which to
found any calculation are but few. If, however, the upper rocks follow the course of
the lower, without the intervention of faults
and irregularities, the tract so occupied would
probably be an oval, forming the centre of the
country, bounded by the sea-coast on the
north and the ridge of primary hills on the
south. From the top of the high land at
Crabb's River, this ridge bounded the horizon,

at the distance apparently of about twenty miles. Allowing half of this width to be occupied by the lower beds, the tract in which it is possible that coal may exist would probably be twenty or thirty miles long, by ten miles wide. Gypsum again appeared once or twice in the cliff, between Crabb's River and St. George's Harbour.

The north side of St. George's Bay, between Cape St. George and Indian Head, is occupied by magnesian limestone, dipping at a slight angle to the north-north-west. Much shale or shaly rock was observed in the cliffs in that direction, north of Cape St. George, but there was some reason to believe that the magnesian limestone lay above the shale. Since I visited that part of the country, I have been told that coal has been found in Port-aux-Ports. I had intended to have examined that place, but having been detained four days off its mouth by continued calms, I gave it up. If coal exist there, it is probable the shale mentioned before is the upper part of the coal formation, and the magnesian limestone may lie over it.*

* There being no previous information to be obtained about this part of the country, it was of course impossible

At Indian Head a mass of igneous rocks appears, but in consequence of the lowness of the land on each side of it, no junction with surrounding formations can be observed. These rocks consist, in one part, of Labrador feldspar, in rather small crystals, aggregated together into a largely granular base, in which are embedded large crystals of bronzite or hypersthene: in another part, they pass into a basaltiform rock, very hard, dark, and heavy, the crystalline parts of which have the cleavage of hypersthene.*

In crossing from St. George's Harbour to the Grand Pond, the country was so covered by diluvial rubbish, and that again by moss and woods, that no guess could be given as to the nature of the rocks below, except from the circumstance of some angular pieces of white limestone being found in one of the brooks. On arriving at the Grand Pond, the cliffs

to say which part of it was most advantageous to examine first. The scattered settlers, north of Cape Ray, generally exhibit a great disinclination to give information and assistance, apparently from a fear of the district being more thickly populated, and taxes and customs introduced.

* I am greatly indebted to Professor Miller of Cambridge, for naming and determining some of the compound minerals in these and other instances.

were found to be gneiss and mica slate. The western end of the island, and the mainland opposite, is chiefly a chlorite schist. All the centre of the island, however, and the main on each side of it, is granite, some of which is white, with mica, some red, with or without hornblende. The northern end of the island is a conglomerate. Immediately opposite the eastern end of the island, on the north shore of the lake, are some thick beds of very white rock, dipping in various directions ; and just beyond these, towards the east, some cliffs of a bright red colour, apparently red sandstone, but the bedding of which is not discoverable. The violence of the winds and waves would not admit of our small boat approaching these cliffs, either in going or returning. Two or three miles east of them, however, the continuation of the same cliffs is composed of beds of red sandstone and marl, passing upwards into brown and yellow sandstones, and conglomerate of small quartz pebbles, interstratified with beds of brown, yellow, and blue marls, clunch, and shales, and dipping at various angles of inclination, but generally moderate ones towards the east,

or south-east. This series of beds is precisely
similar to that previously described as forming
the south side of St. George's Bay, and it
forms the cliffs of all the shores of that part
of the lake north and east of the island. Its
general dip is easterly, and the angle of inclination becomes less as we recede from the
granite and primary rocks. In the bed of a
small brook, at the north-east corner of the
pond, were found various pieces of coal, and
at one part, where the bank was newly fallen,
the following section was exposed :—

	Feet.	Inches.
Sand and boulders . . .	10	0
Softish grey and yellowish sandstone	5	0
Ditto ditto shaly .	1	0
Coal (some part like cannel coal) .	0	6
Yellow clunch	0	2
Grey bind . . .	2	0

All these beds dipped at an angle of 30° to
the south-east. Large pieces of coal were
found in the bed of the brook (which is rapid
and rocky) above this point, showing that
more beds exist, and one Indian of St. George's
Harbour assured me he had seen a bed three
feet thick, in the brook, below this point,
about three years ago. This was probably

true, as I saw many banks in the same brook, where such beds might have appeared, but which were then covered with wood and rubbish that had fallen from above.

It thus appears that the rocks containing beds of coal are those observed to dip towards the wide level tract mentioned before as existing north-east of the Grand Pond, and that as they approach that tract the beds become more horizontal and regular. It is, therefore, highly probable that coal may be found over the whole or greater portion of it. The extent of this low country, however, cannot be very great towards the east, as the land about Red Indian Pond and Hall's Bay is, from all I could learn, high and rugged, and composed of hard slaty rocks, instead of the soft sandstones, clays, and marls of the coal-measures. The head of White Bay is, therefore, the only part of the coast from which this coal-bearing country is at all likely to be easy of access.

The country between Port-aux-ports and the Bay of Islands is lofty and unbroken, and probably occupied chiefly by igneous rocks. Around York and Lark harbours the hills are high, pointed, and precipitous, and consist of igneous rocks of very various characters. In one place a red sienite was seen; in another, not far distant, a dark sienitic rock, containing albite and hypersthene: this dark rock was very abundant, and associated with it was a dark-greenish rock, with dark-red and white veins. It looked, at first sight, like a conglomerate, the veins intersecting each other in every direction, and the pieces enclosed by them being easily detached. It is, however, a trappean rock. Associated with this were masses of a soft crumbly rock, almost made up of little granules, many of which were crystalline carbonate of lime: the whole bore very much the appearance of a peperino, being probably a regenerated volcanic or trappean rock.*

From the neighbourhood of York and Lark harbours, nearly to the head of Humber Sound, the rocks consist of dark brown and red schist or shale, grey gritstones, and black, grey, and red slate. They dip various ways, frequently

* I am much indebted to Professor Sedgwick of Cambridge, as well as Professor Miller, for examining and naming these and other rock specimens.

at high angles, a westerly inclination being
the most frequent. The most irregular contortions occasionally showed themselves, in
which not only were the beds bent, but the
lamination was wrinkled and puckered up
into sharp angles, like a vandyked border.
From this broken condition of its beds, it is
impossible to form an estimate of the total
thickness of the formation with any degree
of accuracy; it occupies, however, the whole
length of the Humber Sound. On approaching the head of the sound, the dip of these
shales and grits becomes more regular, being
always to the west. Their positive junction
with the next formation is nowhere seen, as
the only section, that of the cliffs on the north
side of the Sound, is interrupted by a low
bank of loose sand, 300 or 400 yards across.

Just above this, we come to some beds of
limestone, belonging to a great calcareous
formation, stretching across the mouth of the
Humber River. This limestone is in its
upper portion regularly bedded and variously
coloured, and indistinguishable by mineralogical characters from many secondary lime-

stones. Its lower part, however, is white, crystalline, frequently contains veins and flakes of mica, becomes entangled with quartz rock, and seems to be so intimately associated with the gneiss and mica slate, as to leave no doubt of its belonging to that formation, and being thus entitled to the denomination of a primary limestone. The highest beds of the lime-stone formed a mass about thirty feet thick, of thin shaly beds, of a hard dark grey colour, with brown concretions, that, on a surface which had been some time exposed to the weather, stood out in relief. Below this were seen some thin beds of a hard subcrystalline limestone, some of which was white, and some flesh-coloured, with white veins. This series of beds had a thickness of upwards of 100 feet: they would take a good polish, and would be very ornamental, and, from the thin-ness of some of the beds, would be especially adapted for marble slabs. Below this por-tion lay a few feet of thin-bedded black marble of similar qualities. Then came some grey compact limestone, with thin beds and irregular nodules of whitish chert, which passed down into a large mass of grey compact limestone, in thick beds and without chert, and having a thickness of 300 or 400 feet. In the lower parts of this mass the bedding became indistinct, and it passed down into perfectly white saccharine limestone without any mark of stratification, and but few joints or division-lines of any kind. The hills composed of this limestone are 400 or 500 feet high, and run nearly north and south for a considerable distance. About three miles up the Humber River, its lower portion, in which no appearance of bedding is discernible, forms lofty white precipices of pure marble, crowned and surrounded by thick woods, which, closing in upon the rapids, produce most picturesque scenery. Blocks of any size might here be procured, and by a little management floated down the river into the sound, where any kind of vessel will find excellent anchorage. A little above these limestone precipices, the hills recede from the river, and enclose a valley about two miles in width, but they are continued without any interruption to just below the second rapid, where they close in again on the stream. Here the rocks are gneiss; and mica slate and gneiss form all the hills around the lower end of Deer Pond. At about the middle of this lake, the hills gradually slope down, exposing no cliff: at one point, however, some beds of yellowish sandstone and conglomerate of white quartz pebbles were observed. Round the upper end of Deer Pond, and thence as far as could be seen, was spread the level country mentioned before; but at the rapids just above the bifurcation of the river some ledges of light brown gritstone were seen. Both these gritstones, and the sandstones and conglomerate, were the same rocks as those found on the banks of the Grand Pond and the south side of St. George's Bay, belonging to the lower part of the coal formation. Putting these facts together, we get an east and west section, from the mouth of the Bay of Islands to the head of Grand Pond, which exposes the structure of the country in a satisfactory manner. (See section, No. 13.) The range of hills thus found to be composed, from Cape Ray to the River Humber, of mica slate, gneiss, and their associated rocks, continues to run to the north for many miles, and, as far as I could ascertain, forms an unbroken ridge to Cape Quirpon, the extreme northern point of the island of Newfoundland. In the neighbourhood of Canada Bay, on the western side of White Bay, I have been assured that limestone exists in abundance; and a large specimen which I saw in St. John's, procured from that place, was identical in mineralogical character and appearance with the white marble of the Humber Sound. It is, therefore, highly probable that these hills are composed of mica slate and gneiss, with occasional patches of primary limestone, along the whole west side of the island of Newfoundland.

In Mr. Cormack's journey across the island, from Random Sound to St. George's Bay, he seems to have paid particular attention to the kind of rocks he met with; and though his published notes are scanty, they are in the present instance highly useful. From the hills at the back of St. George's Bay (the Long Range) to the eastern end of Jameson's Lake, nearly in the centre of the island, he mentions no other rock than granite. This accords exactly with the structure of the coast through that space, except that there are also on the coast a few patches of mica slate, gneiss, and other slate rocks. East of Jameson's Lake an abundance of "Serpentine" is mentioned. Mr. Cormack found, north of the Bay of Despair, granite, sienite, quartz, gneiss, fine clay-slate, alum-slate, and indications of coal and iron. The alum-slate was probably shale; the indications of iron might be correct, as that mineral abounds everywhere, but no other indications of coal are to be trusted in such a locality than the exhibition of a bed of that substance itself. Granite and quartz again occurs towards the east; then basalt in flags, as at Belle Isle. If Belle Isle in Conception Bay be meant, it is not basalt, but very fine-grained dark-grey gritstone. There is in this locality probably a patch of the Belle Isle shale and gritstone. Thence to Random Sound nothing is mentioned but granite, mica slate, porphyry, and sienite. It is, therefore, evident that there is no large tract existing in the interior of the country in which the rocks to be found are greatly different from those which come out upon the coast; and as the strike of the rocks throughout Newfoundland, wherever it is not deflected by local accidents, is universally north-north-east and south-south-west, there is clearly no room for any other formations than those already described. All that could be done, then, in carrying out the survey into minuter detail, and examining the interior of the country, would be tracing out the boundaries of the several varieties of igneous rocks and the slate formations that rest upon them, and discovering the obscure relations that exist among the latter. Such operations could obviously be of no public utility, and their benefit to the science of geology is doubtful, and would probably be small, until we are more intimately acquainted with the minutiæ of the structure of the adjacent parts of North America.

There are a few general observations I have to make on the structure of Newfoundland, and the relative age of some of its igneous rocks. The regularity of the strike, and its perfect parallelism throughout the island, is very remarkable. It rarely varies, except for very short distances, where local disturbing forces have affected it, from a true north-north-east and south-south-west course. The only exception is about St. George's Bay, where the rocks strike generally more nearly north-east and south-west, or even in some instances east-north-east and west-south-west. As a consequence of this prevailing strike of the rocks, we find all the other prominent features of the country running in the same direction. Not only do the ranges of hills run in a north-north-east and south-south-west direction, but all the principal lakes, deep bays, and valleys lie in the same line of bearing. The ranges of hills already noticed, the Grand Pond, Red Indian Lake, great part of Gander Pond, Holyrood Pond near St. Mary's Bay, and others; the great inlets of White Bay, Exploits Bay, and Gander Bay; the sounds and inlets of Bonavista Bay, Trinity, Conception, St. Mary's, Placentia, and Fortune Bays, and of course the headlands and peninsulas between all these; the islands of Merasheen, Long Island, and others in Placentia Bay, and the

longitudinal direction of the multitudinous
islands in Bonavista Bay and the Bay of Exploits, all run in a north-north-east and south-
south-west direction. It may also be re-
marked, that in St. George's Bay the line
of the coast on each side strictly coincides
with the strike of the rocks; the south side
running north-east and south-west; the north
side east-north-east and west-south-west. The
Bay of Islands is the only important exception to this general rule throughout the country. To the geologist it may seem something like tautology to speak of the shape and
outlines of the country being parallel to the
strike of the rocks of which it is composed.
To those who are not geologists, however, it
may be interesting to remark how these things
depend upon each other, more especially as
the example afforded by the island of Newfoundland is so clear and striking.

The cleavage of the slate rocks is another
subject deserving of remark. The cleavage of
the slate rocks of Newfoundland is frequently
parallel to the planes of stratification, but
often cuts them at all angles, and is some-

times perpendicular to them. The cleavage is best developed in the rocks of the finest grain, and becomes evanescent invariably on approaching a coarse-grained bed. In beds of the finest grain, however, it is not always present, and is frequently capricious and partial in its appearance. The strike of the cleavage is, in the great majority of instances, parallel to the strike of the beds, but not invariably so. The cleavage is much more constant, as regards its strike and dip in relation to the horizon and the points of the compass, than it is in relation to the strike and dip of the beds, or than these latter are to the horizon and the points of the compass. The dip of the cleavage never forms a less angle with the horizon than 45°, while in the majority of instances it is nearly perpendicular to the horizon. Its strike, when well developed, never, in any instance, where it was carefully observed, was found to vary more than 10° or 15° from a north-north-east and south-south-west bearing. The following list of observations will show the general position of the cleavage, and one or two instances of discordance be

tween the strike of the cleavage and that of
the beds :—

Cleavage of Slate.	Strike.	Dip.
Near St. John's . . .	N. 35° E.	nearly perpendicular.
At Topsail	N. 35° E.	
Aquafort	N. 15° E.	
North Harbour, St. Mary's Bay	N. 15° E.	perpendicular.
*Same place (beds) . .	N. 15° E.	to the West, 45°.
North East Mountain, Placentia	N. 10° E.	Westerly, 80°.
Merasheen Harbour . .	N. 15° E.	Westerly, 85°.
*Same place (beds) . .	N. 50° E.	N.W., 80°.
Indian Harbour, Merasheen Island	N. 15° E.	
Long Harbour, Placentia Bay	N. 20° E.	
Chapel Arm Brook, Trinity Bay	N. 15° E.	perpendicular.
Brigus	N. 10° E.	nearly perpendicular.
Sculpin Island Cove, near Brigus	N. 5° E.	Westerly, 80°.
Spaniard's Bay . . .	N. 20° E.	
Harbour Grace . . .	N. 20° E.	Westerly, 75°.
Harbour Grace Island . .	N. 15° E.	Westerly, 45°.
Carbonear	N. 30° E.	Westerly, 60°.
Catalina Harbour . . .	N. 27° E.	Easterly, 45°.
Clode Sound . . .	N. 20° E.	
Morris's Island, Bonavista Bay	N. 10° E.	Westerly, 45°.
*Ditto (beds) . . .	N.	West, 25°.
Same Island, another part . .	N. 12° E.	perpendicular.
*Ditto (beds) . . .	N.	East, 20°.
Bloody Bay, Bonavista Bay .	N. 25° E.	perpendicular.
Gander Bay . . .	N. 30° E.	
Exploits River . . .	NE. by N.	

*Beds.

It appears, then, that the strike and dip of
the cleavage of the rocks of Newfoundland
are not absolutely dependant on the strike
and dip of the beds, the one often varying
while the other remains constant. But it
appears, also, that the same cause which

gave their *prevalent* and *general* direction to the mechanical forces, by which the rocks were elevated from their original position, and their strike and dip produced, likewise determined the direction in which those forces should act (whatever they were) which produced the cleavage.

As regards the relative age of the igneous rocks, it appears that the granites are generally newer than the mica slate and gneiss, which repose upon them. It is also evident that the large mass of porphyritic granite on the south coast is more modern than some of the shales, flags, and schists about La Poile, inasmuch as these latter are penetrated by veins from the granite. The phenomena seen opposite the western end of Random Island show that the red sienites, which compose much of the country in that direction, are the newest rocks of the neighbourhood, as they partly overlie and certainly have disturbed the Belle Isle shale and gritstone there, which is the most modern *stratified* rock on the east side of the island. The same, likewise, is true of the greenstones of Chapel Arm in Trinity Bay, and of Holyrood in Conception Bay, and there is every probability of the greenstones of the

latter place being of the same age as the mass
of igneous rocks forming the range of hills
between Conception Bay and Renews.   At all
events, these latter rocks are more modern
than the older slate formation, which is cut
through and disturbed by them.   On the other
hand, the red igneous rock (generally a sienite)
forming St. Peter's, and the country between
Cape Chapeau Rouge and Fortune Bay, is in
all probability one of the oldest rocks in the
country, as no veins were observed to proceed
from it into the adjoining formations, and
a rounded pebble of a precisely similar rock
was found in a bed belonging to the older
slate formation in Great Placentia.   A mass of
rocks, too, formed of the detritus, either of these
very igneous rocks or some very similar to
them, exists close in their neighbourhood in
Mortier Bay, in which a pebble of sienite, precisely similar to their general character, was
found imbedded.

From the absence of organic remains, no comparison can be instituted between the age of
the Newfoundland rocks and those of England.
It is highly probable, however, that the coal
formation of Newfoundland is the same as that

of Cape Breton and Nova Scotia, which appears to be nearly or quite contemporaneous with the carboniferous series of western Europe. The coal formation is unquestionably the most modern group of stratified rocks to be found in Newfoundland, and there was certainly an interval between the deposition of the upper and lower slate formations. The mass of the granites and other unstratified rocks are more recent than the lower slate formation; some of them, at least, more recent than the upper slate formation; and they may be more modern even than the coal formation.

Drifted and Superficial Accumulations.

The greater part of Newfoundland is covered by an accumulation of drifted materials, sometimes to the depth of several feet. These, for the most part, consist of coarse rubbly gravel, with patches of sand and clay, and imbedded pieces of rock. In the town of St. John's, this accumulation is well shown in the cuttings of the roads and in digging foundations for houses. It is there found to be

a mass of materials such as those above mentioned, having sometimes a semi-stratified character, the finer parts exhibiting an approach
to horizontal arrangement.   The imbedded
portions of rock are almost wholly slate rock,
and are nearly all angular; some blocks two or
three feet in diameter may be observed, but
the generality are much smaller, and they
are distributed indiscriminately throughout
the gravel, the flat pieces resting at all angles
with the horizon.   On the high lands between
St. John's and Portugal Cove, the drifted
materials have all a similar character, except
that on the higher and more exposed parts
the gravel is not so deep, and sometimes is
absent altogether, and the loose blocks only
are found scattered over the surface.   In
the projecting tongue of land between the
eastern coast and Conception Bay, I never
found any boulders of rock different from the
rocks which compose that district.   The low
land between Topsail and Holyrood, which
consists of the Belle Isle shale formation, is
covered with a vast heap of drifted materials,
which have a greater thickness than usual.
Near Topsail are some huge boulders.   One of

these, lying in a marsh, is twenty-three feet long, twelve feet wide, and twelve feet in height, having probably several more feet buried beneath the surface of the moss. It consisted of a grauwacke conglomerate, precisely the same as is found in the hills about two miles to the east of it. Multitudes of a similar character, but smaller size, were lying about. Indeed, in all the lower lands the great difficulties in the way of forming new roads are the boulders. After the wood has been cut down, and side drains have been formed, the moss shrinks and dries up, and exposes a multitude of blocks of rock, many of which are from one to three feet in diameter, and which are frequently in as great plenty as in the bed of a mountain torrent. This may be seen in the road between the Golds and the Bay of Bulls, or between Topsail and Holyrood, or almost universally in all newly formed roads which cross the lower, and occasionally even the higher grounds.

In the neck of land connecting Ferryland Head with the mainland, a great accumulation of drifted material and boulders is shown, apparently as deep as the cliff is high, which is about fifty feet. In patches in this mass there occurs a fine light-coloured clay, which, after drying and pounding, is used by the inhabitants for plastering and white-washing. Among the numerous boulders on the roadside between Ferryland and Aquafort, I observed some porphyry and sienite; among others, slate rock, gritstone, and conglomerate; and at one point (near the bridge over the brook which forms a cascade into the harbour of Aquafort) lay a large boulder, five or six feet in diameter, of that peculiar rock which looked like a conglomerate of angular pieces of porphyry, and which I mentioned as occurring on the top of the Butterpots Hill, near Renews. This hill may be seen from the rising ground close by, distant about eight or ten miles in a west-south-west direction. This boulder and those composed of sienite and porphyry were much rounded, and some of them quite smooth and polished, though several feet in circumference; the blocks of slate rock, however, were more angular, and occurred in greater abundance, but were generally of smaller size.

In ascending the Butterpots, fragments of

rock without gravel were observed at all heights; and on the very summit, resting on the porphyry, were loose blocks of slate rock and sienite, in some degree rounded, and about two feet in diameter. This is the highest point of the neighbourhood, about 1200 feet above the sea, no other hill being within five miles of it which equals its height within 200 or 300 feet. (See section, No. 4.) The same kind of drifted materials as those now described may be observed over all the country, thence by Trepassée into St. Mary's Bay, except that the sienitic blocks get scarcer as we leave the hills of which the Butterpots forms the southern termination. Blocks of porphyry, however, were seen of considerable size near Peter's River, in St. Mary's Bay. In St. Mary's Harbour, one or two smaller pieces of granite, perfectly rounded, were found on the beach, one of which precisely resembled the granite of La Poile : they might, however, be part of the ballast of a vessel. In the diluvial drift of the neighbourhood were some large blocks of a purple porphyry, but the majority were pieces of slate rock, identical in character with the rocks upon which they reposed,

mingled with boulders of gritstone and conglomerate, some of which resembled the Signal
Hill sandstone, beds of which may probably
exist in the country to the northward. Around
Placentia the boulders are principally slate,
gritstone, and porphyry, being fragments of
the rocks in the neighbourhood. In Merasheen
Harbour, on the contrary, the vast variety
of the materials found as boulders is most
remarkable. Within the space of a few yards,
smooth and rounded blocks were seen of several
hundred pounds in weight, of two varieties
of porphyry, three distinct kinds of granite,
and different kinds of slate rock, grauwacke,
sandstones, gritstones and conglomerates. Here
the majority of the boulders seemed to be of
a different composition from the rocks forming
the hills at the back of the harbour, and to
have been derived from the mainland, and
the islands to the northward. In the brook of
" Come-by-Chance," at the northern extremity
of Placentia Bay, were some large perfectly
rounded blocks of sienite, similar to the sienite
seen opposite the west end of Random Island.
As the chain of hills is continuous between the
two places, the blocks of Come-by-Chance

were probably derived from the hills a few miles westward of that place.

The northern coast of Avalon is as much covered with boulders as the rest of the country, but they seem to consist almost entirely of the rocks of the immediate neighbourhood, as I never observed any granite, sienite, or mica slate, &c., among them. I frequently searched in the finer parts of the gravel about St. John's and other places for shells, or other organic remains, but could never discover any. On the top of Harbour Grace Island there is a thin bed of shingle, covered by about two feet of rather fine gravel, and the lighthouse-keeper assured me he had seen shells in it. I could not find any myself, neither could I certainly ascertain whether the shells seen had not been carried there by sea-birds, as I had often found both seashells and echini lying loose and recently picked of their contents on the top of the cliffs. Over the greater part of the country about Bonavista and the Bay of Exploits the superficial accumulations are abundant. They are of similar character to those already described, but the embedded rocks are more

numerous, and of greater variety, the gravel
being more strictly confined to the valleys and
lower lands. Granite boulders are plentiful,
and frequently of large size. To enter into
details would be useless and tiresome, but a
few of the more striking facts may be mentioned. On the top of the hill, 400 feet high,
at the head of Clode Sound, being the highest
point of the neighbourhood, is a large boulder
of white granite, the hill itself being a red porphyry or sienite, and no white granite known
within twenty or thirty miles. On Man
Point Ridge, in Content Reach, a hill 500
feet high and entirely composed of grey, fine-
grained gritstone and slate rock, many large
boulders of red granite, sienite, and gneiss
were found: some of these were angular and
several feet across. Over the granitic district of Cape Freels no other than granitic
boulders were seen. Fogo Head, rising abruptly from the sea to a height of 400 or
500 feet, and steep on all sides, is composed
of grey gritstone and slate rock, as is all the
adjacent country for four or five miles at
least: on its summit are large boulders of
a red sienite, identical in character with

that found in Hare Bay, five or six miles to
the south.   In Toulinguet, on the road to
Back Harbour, a large block of white granite,
three or four yards in circumference, may be
seen, identical with the white granite on the
east side of the harbour, none of which rock
is found *in situ* on the west.   Over all the
country traversed by the River Exploits the
granitic blocks are large and numerous, the
bed of the river being full of them as far as the
falls, and for some distance above them.   On
the top of Camelin Island, on the south shore,
which consists entirely of red porphyry or
sienite, a large block of grey rock, composed
chiefly of radiated zeolite, was found.   Over
the south side of St. George's Bay, resting on
the soft sandstones and marls of the coal
formation, lie multitudes of blocks of granite,
gneiss, and mica slate, derived in all probability from the hills of the Long Range.   In
the beds of the rivers these were especially
abundant, and I was assured by an inhabitant
of the neighbourhood of Crabb's River, that
immense blocks were brought down by the
ice every spring.   At the entrance of the little
creek of Crabb's River, one block lies in the

sand in mid channel, and any craft drawing more than six feet of water is liable to strike upon it. This was probably brought there by the ice on the breaking up of the river in the spring. From these details no very decided conclusions can be drawn. The blocks seem to have been moved in various directions, and in some instances they have evidently travelled northwards, at least for short distances. All that can be said is, that fragments of rock, frequently of great size, have been removed from their original position in all directions for a few miles; and that where other blocks have been found, apparently derived from a more distant source, that source and its direction are unknown. No diluvial grooves, or scratchings of the surface, were observed, but such markings might easily exist under the general coating of moss, and where the surface was exposed they would probably soon be obliterated by the weather. At the northern end of the Grand Pond, and over the low country beyond, and generally in the valley of the Humber River, there was found a bed of red or yellow sand. This was loose and incoherent, but regularly stratified with lines

of small pebbles. It exhibited everywhere
the same character, and had generally a thickness of twenty or thirty feet. It had evidently been deposited very tranquilly. It is
highly probable that before the Humber had
worn so deep a bed in the rocky channel of
the valley by which it escapes to the sea, the
waters of the Deer Pond stood at a higher
level than they do now, and it certainly is
not improbable that all the low country between and about the Deer Pond and the Grand
Pond may have been one large lake, in the
waters of which the bed of sand now under
consideration was deposited. An obstruction
of twenty or thirty feet high, and sixty or
seventy yards long, in the valley of the Humber, at the upper rapids, would in a few
months again cover much of this low country
with water, and form an immense lake.

In the banks of the River Exploits, from
the falls to the sea, a fine unctuous clay
was observed. It is perfectly plastic, and
would make very fine bricks. It is generally
fifteen or twenty feet thick, lying in thin layers,
usually of a slate colour, but with a reddish
band here and there. It occasionally con-

tains a line of very small pebbles, but no sand, and is found immediately on the brink of the river, and in all parts of the valley where the subjacent rocks do not rise to a greater height than twenty or thirty feet above the present level of the water. Above this clay rests a bed of fine sand, two or three feet thick, passing upwards into a coarse rubbly gravel. Over this comes the diluvial drift of coarse gravel and large boulders. The River Exploits now deposits little or no sediment in its bed, which is full of pebbles of all sizes, with no admixture of sand. At its mouth, however, about the head of the bay, are great banks of soft fine mud, and occasionally higher up the river than the mud are beds and projecting points of sand. It appears, then, from these facts highly probable that the country once stood at a lower level; that the arm of the sea formerly extended much farther up, probably receiving the river at the point which is now the entrance of the ravine at the foot of the falls. The clay bed would then be deposited in the still water below the then mouth of the river, as the mud and clay is now in the still water below the present mouth.

Q 2

As the water became more shallow, the silt, mud, and sand would of course be drifted farther and farther out. If under such circum- stances we suppose an elevation to take place, either gradually or at once, the river would begin to cut back its channel in the hard rocks forming the present ravine below the falls, and after sweeping a channel in the soft mate- rials it had previously deposited, would attack the hard rocks below, forming the present rapids and lower falls. Under no other hypo- thesis will, I think, the formation of the clay in its present valley be intelligible.

I regret that both in this instance, and in that of the sand of the valley of the Humber, my search for organic remains was unsuc- cessful.

Similar facts might be observed in regard to the beds and valleys of several others of the brooks of Newfoundland, more especially Rocky River in St. Mary's Bay, where beds of fine clay also occur. In a country like New- foundland, where nothing like a beach can be seen, except a small pebble bank at the heads of the coves and small bays, raised beaches can hardly be expected. Neither were any lines

of cliffs over or behind the present ones, nor
water lines along their sides, in any case observed. On the north side of St. George's
Bay, however, a little valley near Ship Cove,
running out on the sea, was filled up with a
large mass of diluvial sand and boulders; and
near the top of the low crumbling cliff of these
materials, fragments of shells were found: they
were covered sometimes two or three inches
by clay, were in a rotten, decomposed state,
and were at a height of about thirty feet above
high-water mark. They belong to Mya arenaria, and two or three other species common in other parts of the bay, but no recent
shells could be seen on the beach below, which
was covered with pebbles. The circumstances
were altogether too doubtful to enable one to
say whether the presence of the shells was due
to the agency of birds or the elevation of the
bed of the sea.

PRACTICAL RESULTS.

It is to be regretted that the practical results of this survey are but few, and, in the present state of the colony, rather of a negative than a positive character. They may at least, however, be useful, in restraining rash speculation, if not in encouraging prudent enterprise.

Building Materials.

Stone.—The stratified rocks of the country offer few tolerable building-stones. The gritstones of the Bell Isle shale formation are too hard and splintery to be easily worked, even with the chisel, and although they are often divided by natural joints into regular blocks, the smooth faces of these are not easily bound together by mortar. The Signal Hill sandstone often forms admirable materials for rough work, such as walls, &c.; and the conglomerate beds, in which the pebbles are small and numerous, are so heavy, hard, and durable, as scarcely to be excelled for the construction of

sea-walls, breakwaters, and similar purposes.
Some beds, each about five feet thick, divided
by thin partings of marl, and squared by natural joints into blocks of two or three tons
weight, occur at Flat Rock, in a situation
where their removal would be comparatively
easy. Some of the sandstones of the coal-measures in George's Bay would form good flag-
stones, but I did not see any beds sufficiently
hard and durable to deserve mention for their
building qualities. The limestones of the Humber River, probably those also of Canada Bay,
White Bay, and perhaps at some other points
of the "Long Range" of hills, would make
good, and very handsome, building-stone. It
would be easy to work blocks of any size; and
any shade of colour, from dark grey to white,
might be procured. Ornamental marbles also,
especially marble slabs of a black, grey, mottled-
grey, red, and white colour, and possibly also
blocks of pure white marble fit for the statuary,
might be found, the former sorts in abundance.
Of the unstratified rocks, granite and sienite are
those which are best adapted for building.
The sienite at the head of Conception Bay
would make a very fair building-stone, though
of rather a bright-red colour. Similar stone

occurs in several parts of Placentia Bay, and
is marked on the map by a bright-red colour
with the letter S. The same colour with the
letter G. indicates granite, and wherever that
occurs, good building-stone may be found,
though frequently difficult and expensive to
be procured. Along the west side of Bonavista Bay, in the district of which Greenspond
may be taken as the centre, there is abundance
of excellent granite, some of which cannot be
exceeded in its qualities as a building-stone
either for beauty or durability. In some
places loose blocks of this are now lying on
the surface, as for instance about Chalky Cliff,
near Locker's Bay, Bonavista Bay, and near
Cat Harbour, north of Cape Freels.

Slate.—Very good roofing-slate may be procured in abundance much nearer the capital
and more thickly-peopled parts of the island.
Close to St. John's, in its western outskirts—
as, for instance, at the back of the hospital, or
between that point and Rennie's mill, or anywhere along the same line of bearing, and
in the banks of the brook above Waterford
bridge—very good slate rock shows itself, which,
though rotten and brittle at the surface, would
in all probability be of superior quality below.

Some few trials must of course be made, and some little money expended perhaps, in the search, before the best spots are hit upon, as the cleavage in all slate rocks (on which their goodness as roofing-slate depends) is a little capricious and uncertain. On the west side of Conception Bay, from Brigus to Carbonear, and probably farther to the north, no trials are needed, as the slate rock is exposed in the cliffs, and the best spots are obvious to the most inexperienced eye. Harbour Grace Island, especially, is one mass of the most excellent roofing-slate, where slabs of any size might be procured, and split to any required degree of thinness. It must always be borne in mind that the rottenness of the present exposed surfaces is due to the action of the weather. Where-ever the colour of the map indicates the occurrence of beds of the inferior slate-formation, roofing-slate may be found in all probability somewhere in the neighbourhood, should it be considered worth the search.

MINERALS IN BEDS.

Coal.—The beds of coal on the south side of St. George's Bay, as well as in the coun-

try north of the Grand Pond, do not seem
to be of any great thickness. It is perfectly
possible, however, that more important beds
may be found, should the districts ever be
thought worth working. This can only become the case, either from the exhaustion of
the present mines of Cape Breton, or from the
settlement and increased population of the districts themselves. In no other part of the island
of Newfoundland can coal ever be found. This
will at once be obvious to the geologist from
the mere inspection of the map : to others the
reasons will be apparent when they consider
that coal does not occur in detached veins or
masses, or appear indiscriminately in all rocks,
but that it is an integral part of one particular
formation* of great thickness and extent, which
wherever it occurs must occupy a considerable
tract of country. As the beds of coal bear but
a very small proportion to the whole mass of
the formation, large tracts of country might be
occupied by part of the formation in which
there was no coal, but it would be highly im-

* This assertion, of course, is not meant to apply universally, as in some countries coal is found in formations different from those which contain it in others.

probable that beds of coal should exist without some considerable part of the formation to which they belong.

Gypsum.—The gypsum of Codroy Harbour, and in several parts of the south side of St. George's Bay, might very easily be extracted and shipped in large quantities, as it comes out in abundance on the sea-cliffs. As it forms part of the coal formation, the same reasoning will apply to the extension of these beds as to those of the coal.*

Lime.—Besides the beds of limestone mentioned above, a few thin beds of very inferior quality are found in Mortier Bay and Chapel Cove, in Holyrood, and in Conception Bay. In the former place some beds of marl also occur, and calcareous concretions are occasionally found in the variegated slate rocks, but rarely in sufficient quantity to be useful. The calcareous nature of that formation, however, increases its value in an agricultural point of view.

MINERALS IN VEINS.

In the Signal Hill sandstone of Shoal Bay,

* Gypsum occurs in various formations, but very rarely if ever in hard slaty rocks, such as are found in the remainder of Newfoundland.

south of Petty Harbour, is a small vein containing crystals of sulphuret and green carbonate of copper. It was worked during the middle of the last century to a slight extent, and some attempts to renew the work have lately been made, but, I fear, without success. It appears to be a small and irregular vein, without any band of ore, having small nests and strings of the minerals above mentioned disseminated in the stone about it. As this stone is intensely hard, it does not appear that it could be worked to profit, unless the ore were much more abundant than it has been found at present to be. On the western side of the Harbour of Great St. Lawrence, a small vein or string was seen in the sienite, containing crystals of galena or lead-ore, and fluate of lime: they were very trifling, and did not promise to lead to anything more abundant.

These are the sole examples I have met with of mineral veins; and in a country where bare sea-cliffs are so abundant, it does not appear likely that such veins can exist (near the sea at least) without being discovered.

In the interior of the country, search for such uncertain things would be endless, and should they exist, their discovery must be left to chance.

Agricultural Characters.

Coal Formation.—The materials of which this is chiefly composed, soft sandstones and rich marls, generally form very fertile districts. Accordingly, in the tracts occupied by this formation in Newfoundland grass will grow without cultivation, and the timber is of a somewhat different and better character than that generally met with. The thick coating of moss, however (the curse of the country), spreads even here, and it is only in the small spots cleared by the few settlers of the south side of St. George's Bay that the superiority of the soil is strikingly manifest.

Upper Slate Formation.—The Agricultural character of both portions of this formation, namely, the Bell Isle shales and the variegated slates, is superior to that of the lower rocks. Bell Isle itself, and the small spots about the points of the headlands at the bottom of Conception Bay, are instances of this. The same thing may be inferred from the timber of Witless Bay, and the natural grass and clover found occasionally along the east shore of Trinity Bay, from Heart's Desire to Chapel Arm.

Wherever the colours of this formation appear on the map, the soil may be safely taken as above the average Newfoundland value.

Inferior Slate Formation.—The upper part of this, namely, the Signal Hill sandstone, is sterile in the extreme. The St. John's slate rock varies greatly, and the quality of its soil depends much on its situation, slope, and drainage, as also on the kind of gravel by which it is covered : the valleys of the formation often afford good garden-ground.

Gneiss and Mica Slate.—This is generally utterly sterile and desert. Where, however, it contains limestone, as in the Humber River, its character is greatly altered, and the timber growing upon it is found to excel that of the rest of the island both in size and quality.

The igneous rocks are generally hard, and, where not covered by gravel or other débris, are hopelessly barren. This character may be stated, with very few exceptions, to attend all those districts coloured bright red in the map. The districts in which tolerable timber may yet be procured are, the Bay of Islands, the neighbourhood of the Grand Pond, some parts of the south side of St. George's Bay, the neighbourhood of Rocky River, and the head of St. Mary's Bay, Witless Bay and Random Sounds in Trinity Bay, the country on the south-west side of Bonavista Bay, from Goose Bay to Freshwater Bay,* the Bay of Exploits, about the head and along the banks of the river a mile or two from the sea, and, I believe, also Green Bay and White Bay.

In speaking of the agricultural characters of the different formations, it must be borne in mind that, after all, a low and sheltered situation (in the climate of Newfoundland), good natural or artificial drainage, and the kind of gravel and other detritus interposed between the rock and the surface exercise the most important influence on the relative fertility of different spots. Anything approaching to rich soil can only be found in the alluvial strips of land about the larger brooks and rivers, and there only in scattered spots. The country is generally entirely destitute of vegetable mould, and can never, therefore, under any circumstances, become an extensively agricultural one. There

* All the good timber within a mile or two from the shore of Bonavista Bay has been already exhausted, but large woods stretch into the interior.

is at the same time no doubt, that were roads
opened between the richer and more popu-
lous districts (as between the different bays of
Avalon and St. John's) quite enough beef,
mutton, and vegetables might be produced to
supply the wants of the population. For the
production of this supply, the upper parts of
St. Mary's and Trinity Bays are the most pro-
mising.

ROADS.

On a subject so intimately connected with
the present one, as the construction of roads, I
may perhaps be pardoned, and not considered
officious, if I offer a few observations.

After the completion of the proposed line
from the head of Holyrood in Conception
Bay to Salmonier, and thence by Colinet to
Placentia, it appears to me that a road along
the valley of the Rocky River, branching at
the bifurcation of that river, and continued to
Dildo Cove in Trinity Bay on the one side,
and along the bank of the Hodge River to
Snow's Pond, and through the valley of the
Golds to Port-de-Grave on the other, would
be a very advantageous line. By keeping

along the valleys, the roads would be naturally level, cheaply constructed, and, besides these advantages in a commercial point of view, would lay open the most sheltered and thickly-wooded portion of Avalon, that in which the timber is of the greatest size, and the soil probably of the best quality. The eastern bank of the river seems the most level and accessible.

In some of the new lines of road everything seems to have been sacrificed to straightness of direction. It is evident that a straight road can only be the best when traversing a plain. In a hilly country a straight road is sure to be the most difficult, tiresome, and expensive, and may even be longer than some other line that can be found. For in a steep hill the curve of the road over the hill may possibly be as long as the curve round its side, while there is no comparison in the labour, either of construction or conveyance. I met with two or three instances, in which short and sudden pitches, rendering the road quite impracticable for wheel-conveyances of any kind, without much cutting away and filling up, might have been easily avoided by a gentle deflection of the road for a few yards. One or two hills of considerable length and elevation were in like manner uselessly surmounted.

In laying out a road, moreover, round a harbour, the great advantage seems to have been neglected of taking the road at some little distance from the beach, so as to secure three frontages, namely, the two sides of the road and the beach, instead of merely removing one, namely, the beach, a few feet farther back. The present convenience of individuals ought surely to give way before the prospective advantage of the whole community.

In taking a road across a marsh, I observed that the simple expedient of laying a matting of boughs and branches across the road, before putting on the gravel, was frequently neglected. I can speak, from actual inspection, of the sound and durable nature of roads thus constructed in other localities.

THE END.

London: Printed by WILLIAM CLOWES and SONS, Stamford-street.

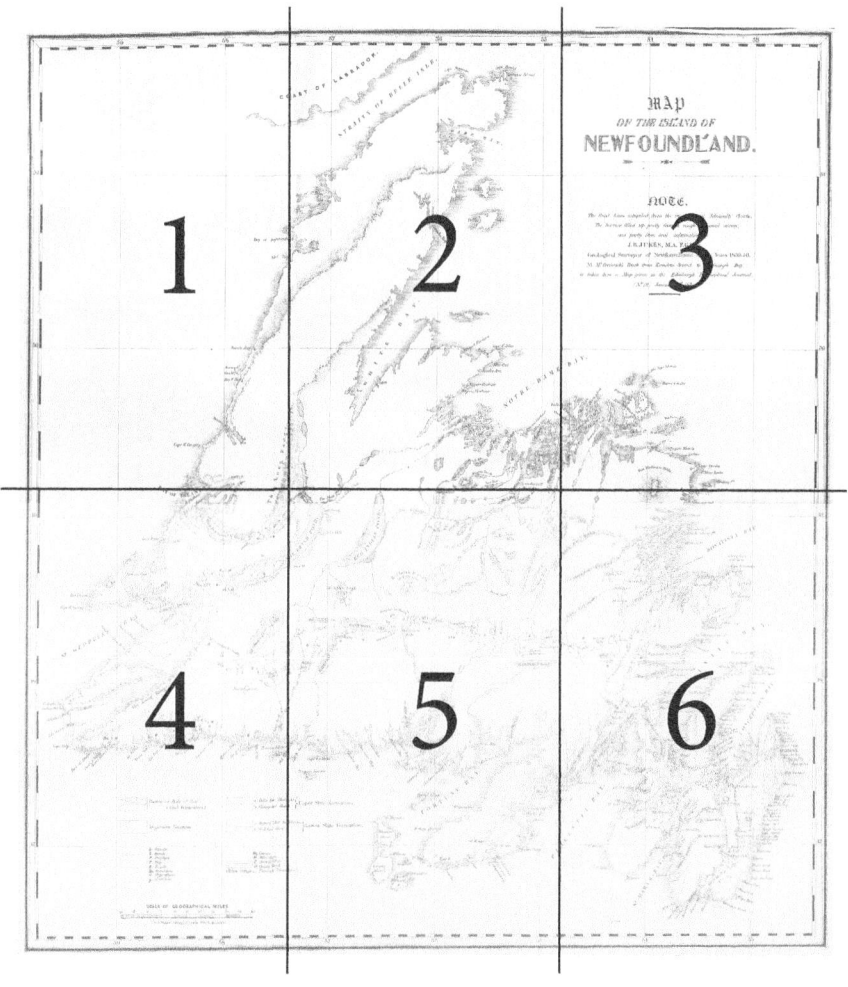

Key to Following Pages

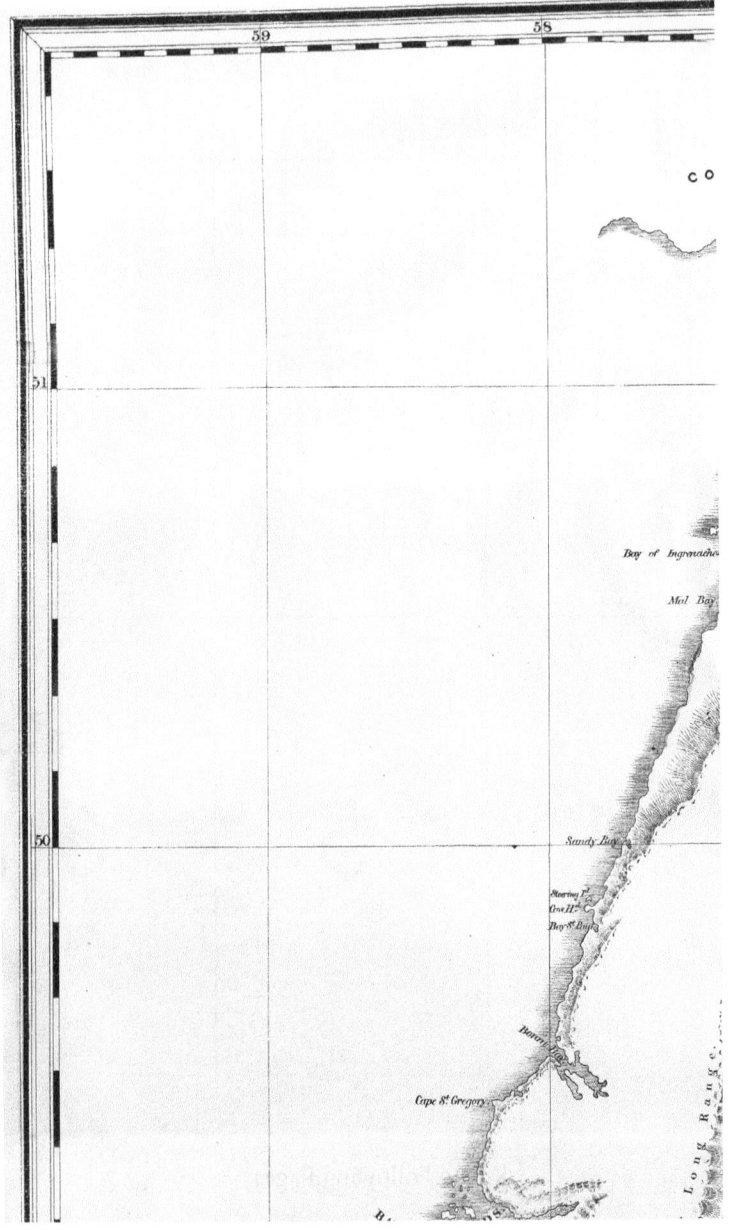

1

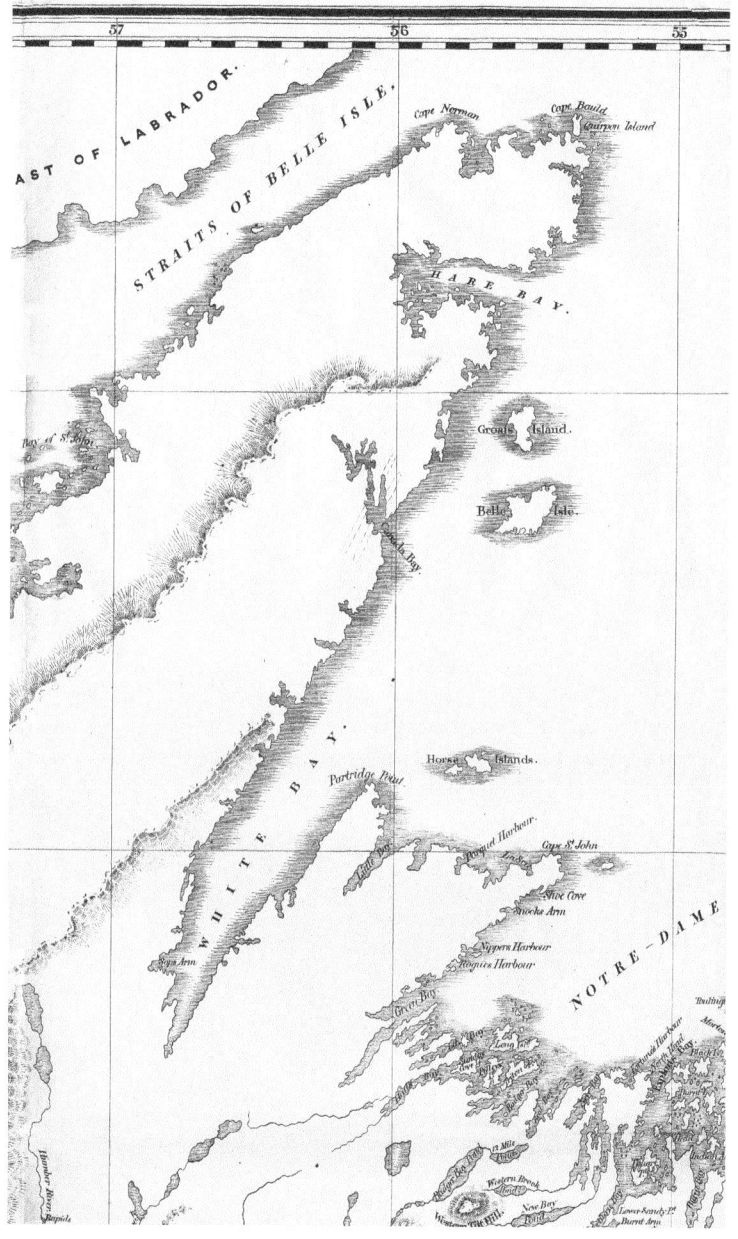

57 56 55

COAST OF LABRADOR.

STRAITS OF BELLE ISLE.

Cape Norman   Cape Bauld
Quirpon Island

HARE   BAY.

Bay of St Sho.

Groais   Island.

Belle   Isle.

Canada Bay.

Horse   Islands.

Partridge Point.

Sop Arm.

Fichot Bay.

WHITE   BAY.

Brequet Harbour.

Cape St John.

Shoe Cove.

Snooks Arm.

Nippers Harbour.

Hognies Harbour.

NOTRE-DAME

Green Bay.

Long Jack

Halfway

Morte

Western Harbour

Harbour Deep

Rapids

17 Mile

Western Brook.

New Bay.

Lower Sandy Pt.

Burnt Arm.

2

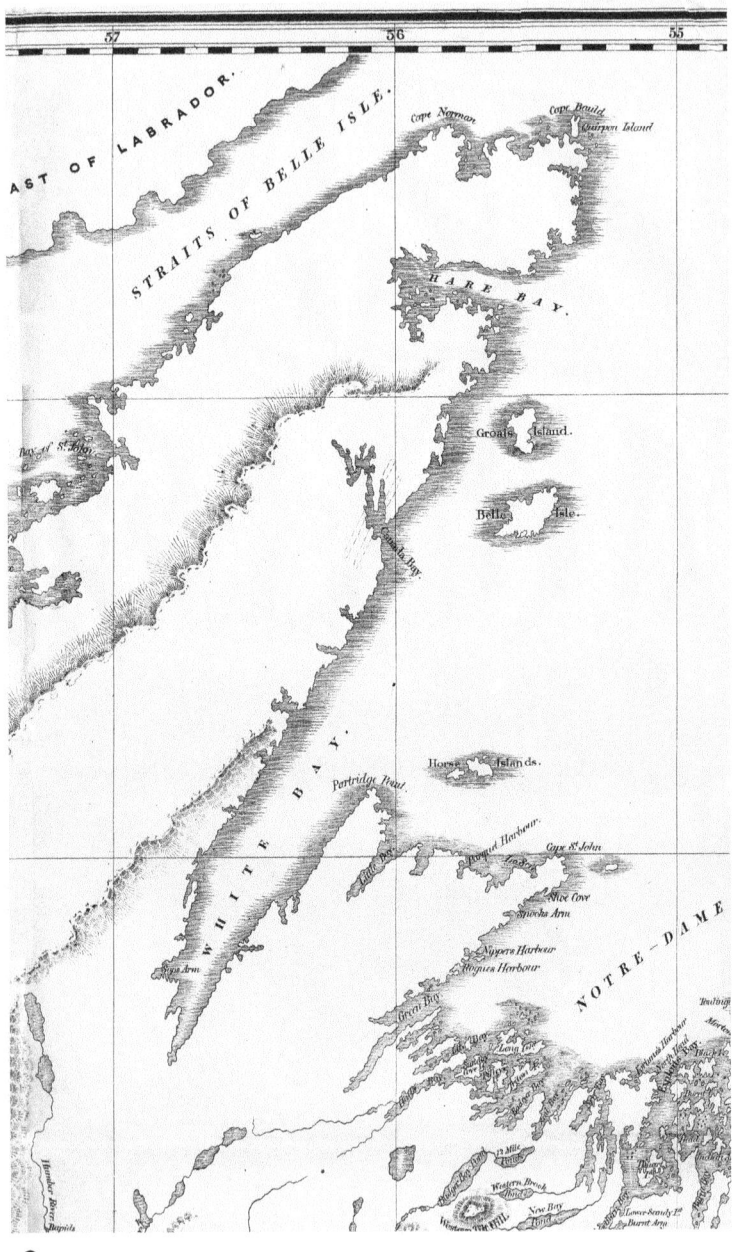

COAST OF LABRADOR.

STRAITS OF BELLE ISLE.

Cape Norman

Cape Bauld
Quirpon Island

HARE BAY.

Bay of St John

Groais Island.

Belle Isle.

Canada Bay

WHITE BAY.

Partridge Road.

Horse Islands.

Sops Arm

Little Bay

Round Harbour

Cape St John

Aloé Cove

Snooks Arm

Nippers Harbour

Bognies Harbour

NOTRE-DAME

Green Bay

Long Pond

Western Brook
Head

New Bay
Head

Lower Sandy Pt
Big Burnt Arm

2

𝕸𝖆𝖕
OF THE ISLAND OF
NEWFOUNDLAND.

➤➤ ✦ ⇐

51

𝕹𝖔𝖙𝖊.

The Coast Lines compiled from the most recent Admiralty Charts;
The Interior filled up partly from a rough personal survey;
and partly from oral information; by

J.B.JUKES, M.A. F.G.S.&c.

Geological Surveyor of Newfoundland in the Years 1839-40.

N.B. M.ʳ Cormack's Track from Random Sound to S.ᵗ George's Bay;
is taken from a Map given in the Edinburgh Philosophical Journal,
(N.ᵒ 19.) January 1.ˢᵗ 1824.

50

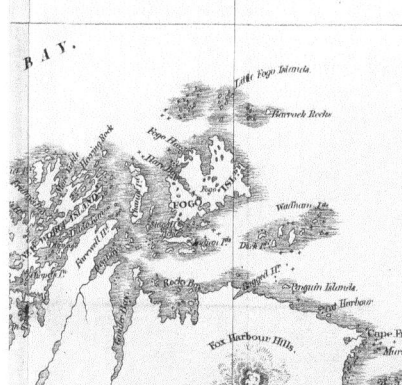

BAY.

Little Fogo Islands

Barrack Rocks

Fogo Harbour

FOGO ISLAND

Wadham I.ᵈˢ

Dog I.ᵈ

Rocky Bay

Penguin Islands

Ragged Harbour

Fox Harbour Hills

Cape Freels.

Murr Rocks

Stinking I.ᵈˢ

Thistle I.

Flowers P.ᵗ

3

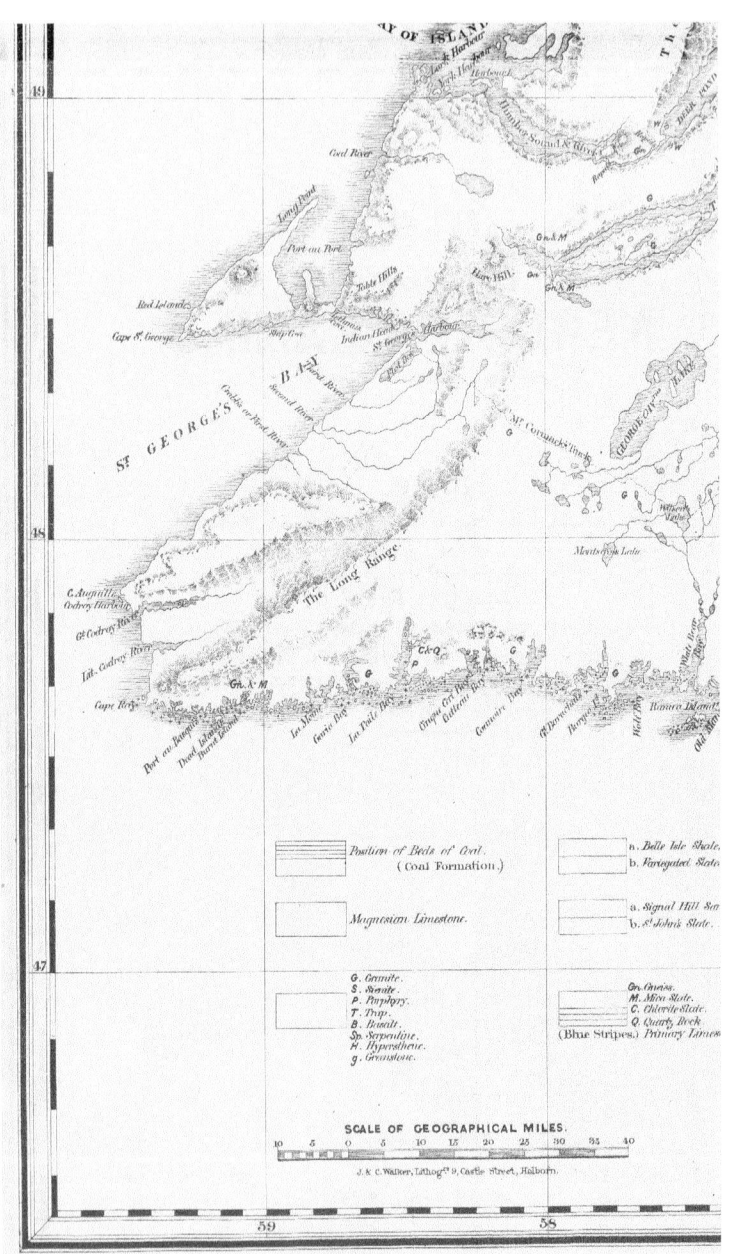

SCALE OF GEOGRAPHICAL MILES.

10 5 0 5 10 15 20 25 30 35 40

J. & C. Walker, Lithog.rs 9, Castle Street, Holborn.

4

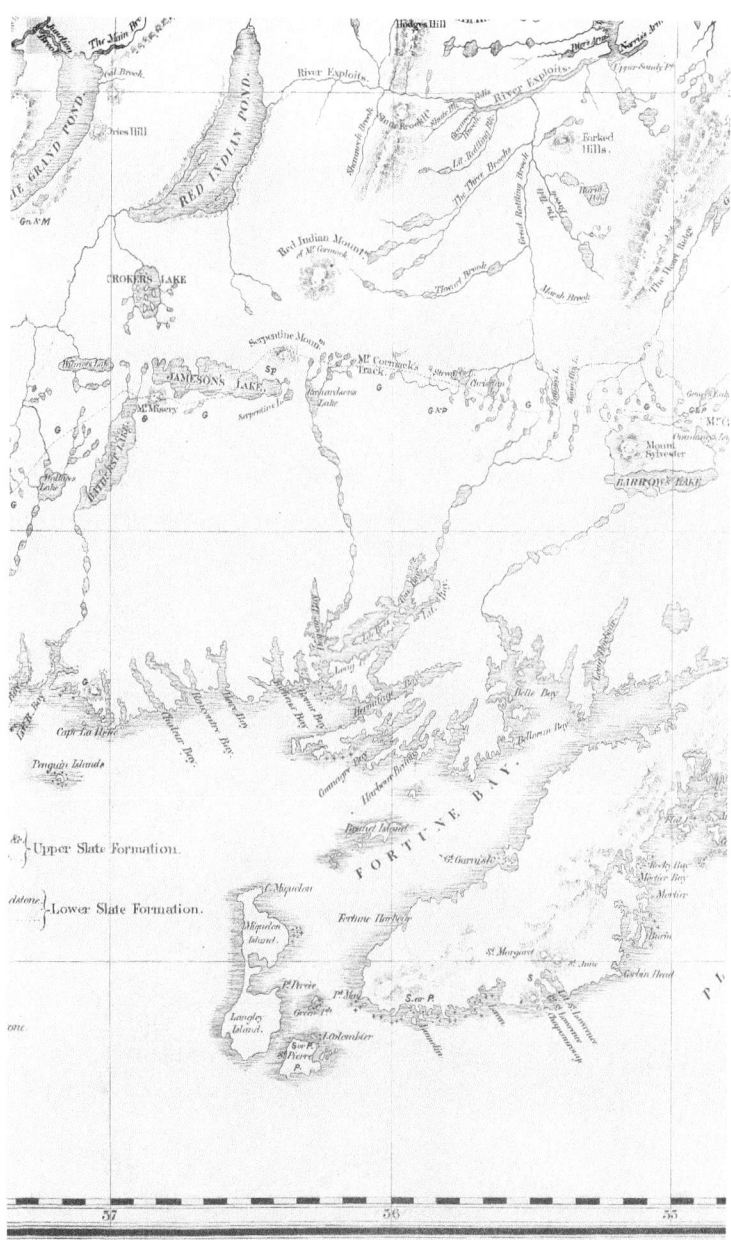

5

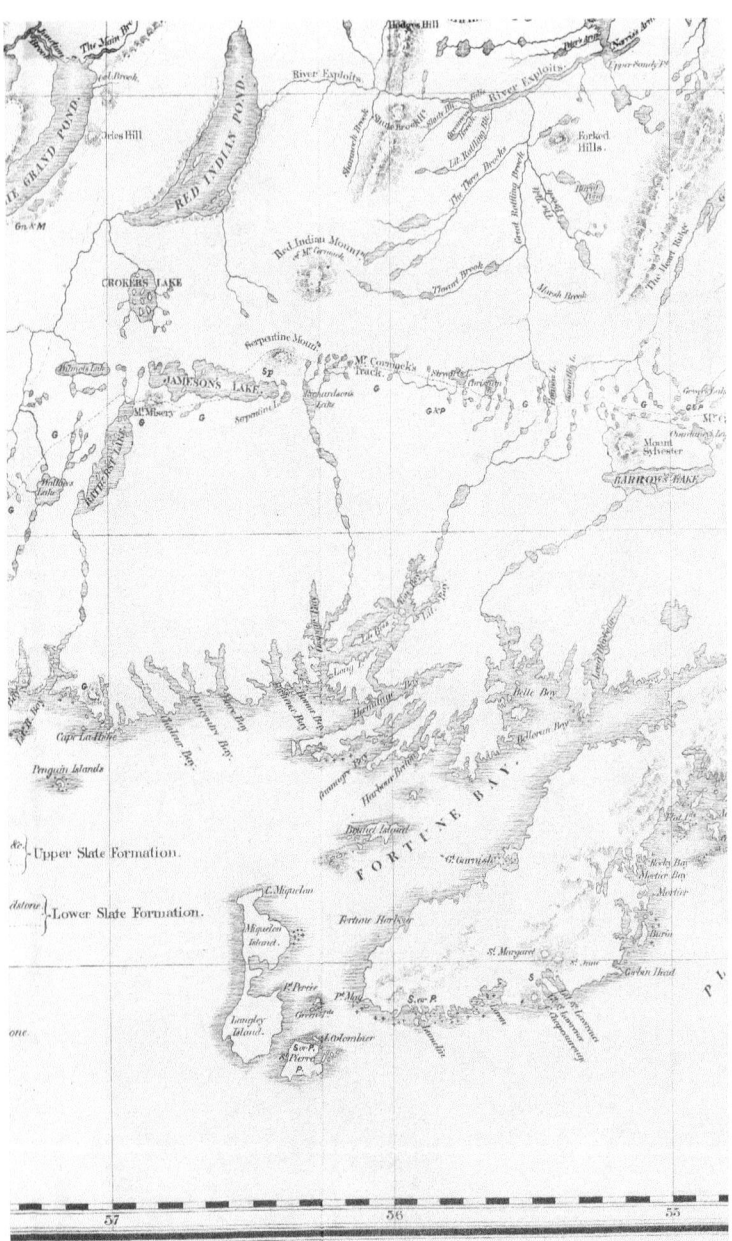

Upper Slate Formation.

Lower Slate Formation.

5

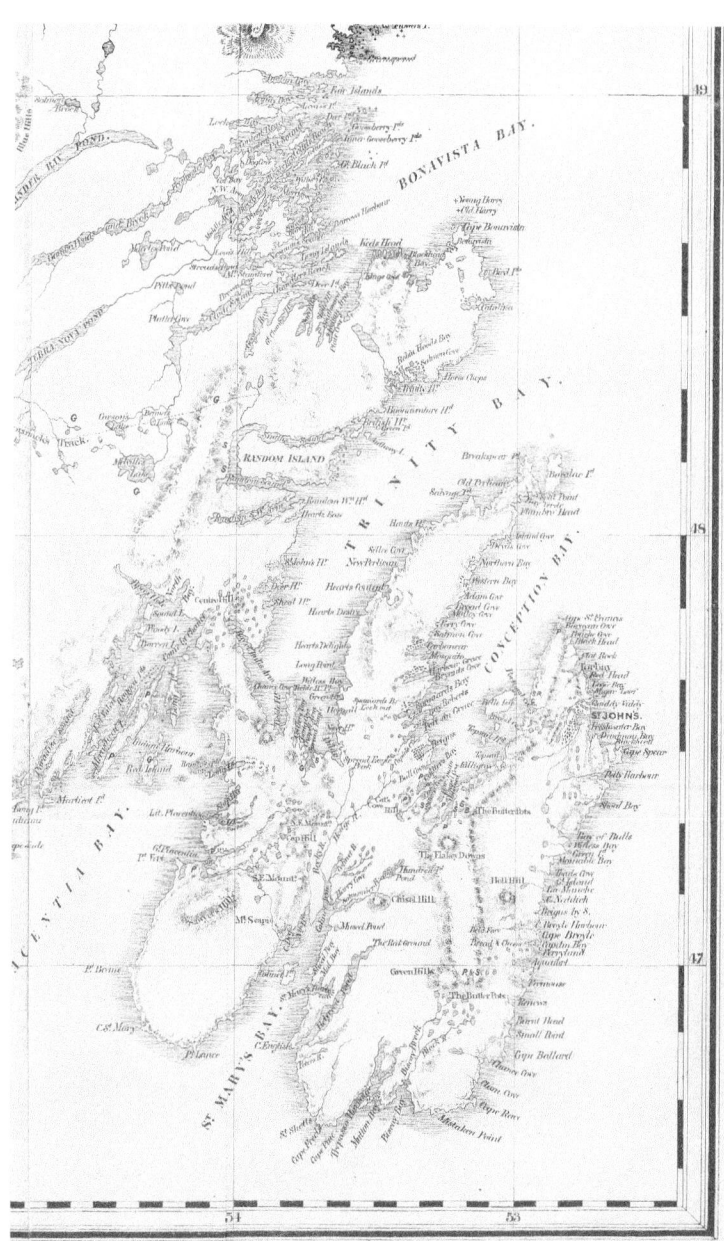

6